Marine Microorganisms

Extraction and Analysis of Bioactive Compounds

Food Analysis & Properties

Series Editor
Leo M. L. Nollet
University College Ghent, Belgium

Marine Microorganisms: Extraction and Analysis of
Bioactive Compounds
Edited by Leo M. L. Nollet

(2016)

Flow Injection Analysis of Food Additives
Edited by Claudia Ruiz-Capillas and Leo M. L. Nollet

(2015)

Food Analysis & Properties Series

Marine Microorganisms

Extraction and Analysis of Bioactive Compounds

EDITED BY LEO M. L. NOLLET

CRC Press
Taylor & Francis Group
Boca Raton London New York

CRC Press is an imprint of the
Taylor & Francis Group, an **informa** business

CRC Press
Taylor & Francis Group
6000 Broken Sound Parkway NW, Suite 300
Boca Raton, FL 33487-2742

© 2017 by Taylor & Francis Group, LLC
CRC Press is an imprint of Taylor & Francis Group, an Informa business

No claim to original U.S. Government works

Printed on acid-free paper
Version Date: 20160414

International Standard Book Number-13: 978-1-4987-0255-3 (Hardback)

Library of Congress Cataloging-in-Publication Data

Names: Nollet, Leo M. L., 1948- , editor.
Title: Marine microorganisms : extraction and analysis of bioactive compounds
/ [edited by] Leo M.L. Nollet.
Other titles: Food analysis and properties.
Description: Boca Raton : CRC Press/Taylor & Francis, 2017. | Series: Food
analysis and properties | Includes bibliographical references and index.
Identifiers: LCCN 2016013144 | ISBN 9781498702553 (hardback : alk. paper)
Subjects: | MESH: Biological Products--analysis | Aquatic Organisms | Dietary
Supplements--analysis | Functional Food
Classification: LCC QH90 | NLM QV 241 | DDC 578.76--dc23
LC record available at http://lccn.loc.gov/2016013144

Visit the Taylor & Francis Web site at
http://www.taylorandfrancis.com

and the CRC Press Web site at
http://www.crcpress.com

Printed and bound in the United States of America by Publishers Graphics,
LLC on sustainably sourced paper.

Contents

Series Preface   vii

Preface   ix

Editor   xi

Contributors   xiii

SECTION I   BIOACTIVE COMPOUNDS AND MARINE MICROORGANISMS

Chapter 1   Bioactive Marine Natural Products: Insights into Marine
            Microbes, Seaweeds, and Marine Sponges as Potential
            Sources of Drug Discovery   3
            *Joseph Selvin, A. S. Ninawe, G. Seghal Kiran, S. Arya,
            and S. Priyadharshini*

SECTION II   NUTRACEUTICALS

Chapter 2   Isolation and Screening for Bioactive Compounds   19
            Rama Bhadekar, Anuradha Mulik, and Sonali Ambulkar

Chapter 3   Extraction, Composition, and Quantification of
            Carbohydrates   35
            Ompal Singh, H. S. Rathore, and Leo M. L. Nollet

Chapter 4   Understanding the Role of Cell Disruption Methods in
            Extracting Lipids   61
            Avinesh R. Byreddy and Munish Puri

Chapter 5   Protein Nutraceuticals from Marine Microbes   75
            *Lipsy Chopra, Gurdeep Singh, Ramita Taggar, Raj Kumar,
            and Debendra K. Sahoo*

Chapter 6   Microalgae as a Source of Pigments: Extraction and
               Purification Methods   99
               Helena M. Amaro, Isabel Sousa-Pinto, F. Xavier Malcata,
               and A. Catarina Guedes

Chapter 7   Opening Avenues in Marine Probiotics: Present and Future   129
               Ira Bhatnagar, Mani Vasagan, and P. V. Bramhachari

Index   147

Series Preface

There will always be a need for analyzing methods of food compounds and properties. Current trends in analyzing methods include automation, increasing the speed of analyses, and miniaturization. The unit of detection has evolved over the years from micrograms to picograms.

A classical pathway of analysis is sampling, sample preparation, cleanup, derivatization, separation, and detection. At every step, researchers are working and developing new methodologies. A large number of papers are published every year on all facets of analysis. So, there is a need for books that gather information on one kind of analysis technique or on analysis methods of a specific group of food components.

The scope of the CRC Series on Food Analysis & Properties aims to present a range of books edited by distinguished scientists and researchers who have significant experience in scientific pursuits and critical analysis. This series is designed to provide state-of-the-art coverage on topics such as

1. Recent analysis techniques on a range of food components
2. Developments and evolution in analysis techniques related to food
3. Recent trends in analysis techniques of specific food components and/or a group of related food components
4. The understanding of physical, chemical, and functional properties of foods

The book *Marine Microorganisms: Extraction and Analysis of Bioactive Compounds* is the second volume of this series.

I am happy to be a series editor of such books for the following reasons:

- I am able to pass on my experience in editing high-quality books related to food.
- I get to know colleagues from all over the world more personally.
- I continue to learn about interesting developments in food analysis.

A lot of work is involved in the preparation of a book. I have been assisted and supported by a number of people, all of whom I would like to thank. I would especially like to thank the team at CRC Press/Taylor & Francis, with a special word of thanks to Steve Zollo, senior editor.

Many, many thanks to all the editors and authors of this volume and future volumes.

I very much appreciate all their effort, time, and willingness to do a great job. I dedicate this series to

- My wife, for her patience with me (and all the time I spend on my computer)
- All patients suffering from prostate cancer; knowing what this means, I am hoping they will have some relief

Dr. Leo M. L. Nollet (Retired)
University College Ghent
Ghent, Belgium

Preface

Almost three-fourths of the Earth is occupied with seawaters, which make them an unmatched resource for biologically potential materials that could prove beneficial for human health. Among the various kinds of life forms that exist in the sea, marine flora occupies the top position. Among the marine flora, marine microbes have a prominent occurrence and purpose in this area.

The bioactives from marine microorganisms include antibiotic compounds, polysaccharides, inhibitors, enzymes, peptides, and pigments. These are used in various fields of biology that range from nutraceuticals to cosmeceuticals. Recent scientific investigations have revealed that marine microbial compounds exhibit various beneficial biological effects such as anti-inflammatory, anticancer, anti-HIV, antihypertensive, antidiabetic, and several more medicinal effects. Thus, there is a greater need for knowledge and understanding the importance of the bioactive components that can be gained from marine microorganisms.

The major focus of this book is to shed light on the extraction, cleanup, and detection methods of major compounds from marine organisms. The book consists of seven chapters which are divided into two sections. Section I provides insight on bioactive marine natural products: marine microbes, seaweeds, and marine sponges as potential sources of drug discovery. Section II focuses on the analysis methods of the biocomponents from marine microorganisms—these topics can assist researchers or students to gain an understanding about current isolation and analysis methods of the bioactives.

This book aims to provide an idea about the various bioactives of marine microbes (bacteria, fungi, diatoms, and others) toward nutraceutical (and pharmaceutical) development from preliminary research to the final product.

Most chapters deal with the extraction and analysis of such bioactives in marine microbes and final products.

I have the pleasant duty to thank all the contributors of this work. Their enthusiasm has resulted in this small but excellent book.

I dedicate this volume to my five grandchildren: Fara, Fleur, Kato, Naut, and Roel.

Keep reading books, but remember that a book is only a book, and you should learn to think for yourself.

—Maxim Gorky

Editor

Leo M. L. Nollet, PhD, received an MS (1973) and PhD (1978) in biology from the Katholieke Universiteit Leuven, Belgium.

He is an editor and associate editor of numerous books. He edited for Marcel Dekker, New York—now CRC Press/Taylor & Francis—the first, second, and third editions of the books entitled *Food Analysis by HPLC* and the *Handbook of Food Analysis*. The last edition is a two-volume book. He also edited the *Handbook of Water Analysis* (first, second, and third editions) and *Chromatographic Analysis of the Environment*, Third Edition (CRC Press).

With F. Toldrá, he coedited two books published in 2006 and 2007: *Advanced Technologies for Meat Processing* (CRC Press) and *Advances in Food Diagnostics* (Blackwell Publishing—now Wiley). With M. Poschl, he coedited the book *Radionuclide Concentrations in Foods and the Environment* also published in 2006 (CRC Press).

Dr. Nollet has also coedited several books with Y. H. Hui and other colleagues: the *Handbook of Food Product Manufacturing* (Wiley, 2007), the *Handbook of Food Science, Technology and Engineering* (CRC Press, 2005), *Food Biochemistry and Food Processing* (first and second editions; Blackwell Publishing—now Wiley, 2006 and 2012), and the *Handbook of Fruits and Vegetable Flavors* (Wiley, 2010).

In addition, he edited the *Handbook of Meat, Poultry and Seafood Quality* (first and second editions; Blackwell Publishing—now Wiley, 2007 and 2012).

From 2008 to 2011, he published with F. Toldrá five volumes on animal product–related books, namely, the *Handbook of Muscle Foods Analysis*, the *Handbook of Processed Meats and Poultry Analysis*, the *Handbook of Seafood and Seafood Products Analysis*, the *Handbook of Dairy Foods Analysis*, and the *Handbook of Analysis of Edible Animal By-Products*. Also in 2011 with F. Toldrá, he coedited for CRC Press two volumes: *Safety Analysis of Foods of Animal Origin* and *Sensory Analysis of Foods of Animal Origin*. In 2012, they both published the *Handbook of Analysis of Active Compounds in Functional Foods*.

In a coedition with Hamir Rathore, the book *Handbook of Pesticides: Methods of Pesticides Residues Analysis* was marketed in 2009; *Pesticides: Evaluation of Environmental Pollution* in 2012; and the *Biopesticides Handbook* in 2015.

Other finished book projects include *Food Allergens: Analysis, Instrumentation, and Methods* (with A. van Hengel; CRC Press, 2011) and *Analysis of Endocrine Compounds in Food* (Wiley-Blackwell, 2011).

Dr. Nollet's recent projects include *Proteomics in Foods* with F. Toldrá (Springer, 2013) and *Transformation Products of Emerging Contaminants in the Environment: Analysis, Processes, Occurrence, Effects and Risks* with D. Lambropoulou (Wiley, 2014).

In this series, CRC Food Analysis & Properties, he edited with C. Ruiz-Capillas, *Flow Injection Analysis of Food Additives* (CRC Press, 2015).

Contributors

Helena M. Amaro
Interdisciplinary Centre of Marine and
    Environmental Research
University of Porto
and
Institute of Biomedical Sciences Abel
    Sazar
Porto, Portugal

Sonali Ambulkar
Department of Microbial Biotechnology
Rajiv Gandhi Institute of IT and
    Biotechnology
Bharati Vidyapeeth University
Pune, India

S. Arya
Department of Food Science and
    Technology
Pondicherry University
Puducherry, India

Rama Bhadekar
Department of Microbial Biotechnology
Rajiv Gandhi Institute of IT and
    Biotechnology
Bharati Vidyapeeth University
Pune, India

Ira Bhatnagar
Centre for Cellular and Molecular Biology
Hyderabad, India

P. V. Bramhachari
Department of Biotechnology and Botany
Krishna University
Machilipatnam, India

Avinesh R. Byreddy
Centre for Chemistry and Biotechnology
School of Life and Environmental Sciences
Deakin University
Geelong, Victoria, Australia

Lipsy Chopra
Institute of Microbial Technology
Council of Scientific and Industrial
    Research
Chandigarh, India

A. Catarina Guedes
Interdisciplinary Centre of Marine and
    Environmental Research
University of Porto
and
Laboratory of Process Engineering,
    Biotechnology and Energy
Porto, Portugal

G. Seghal Kiran
Department of Food Science and
    Technology
Pondicherry University
Puducherry, India

Raj Kumar
Institute of Microbial Technology
Council of Scientific and Industrial
    Research
Chandigarh, India

F. Xavier Malcata
Department of Chemical Engineering
University of Porto
and
Laboratory of Process Engineering,
    Biotechnology and Energy
Porto, Portugal

Anuradha Mulik
Department of Microbial Biotechnology
Rajiv Gandhi Institute of IT and
    Biotechnology
Bharati Vidyapeeth University
Pune, India

A. S. Ninawe
Department of Biotechnology
Ministry of Science and Technology
New Delhi, India

S. Priyadharshini
Department of Food Science and
    Technology
Pondicherry University
Puducherry, India

Munish Puri
Bioprocessing Laboratory
Centre for Chemistry and Biotechnology
Deakin University
Geelong, Victoria, Australia

H. S. Rathore
Department of Applied Chemistry
Aligarh Muslim University
Aligarh, India

Debendra K. Sahoo
Institute of Microbial Technology
Council of Scientific and Industrial
    Research
Chandigarh, India

Joseph Selvin
Department of Microbiology
School of Life Sciences
Pondicherry University
Puducherry, India

Gurdeep Singh
Institute of Microbial Technology
Council of Scientific and Industrial
    Research
Chandigarh, India

Ompal Singh
Department of Research in Unani
    Medicine
Aligarh Muslim University
Aligarh, India

Isabel Sousa-Pinto
Interdisciplinary Centre of Marine and
    Environmental Research
University of Porto
and
Faculty of Sciences
Department of Biology
University of Porto
Porto, Portugal

Ramita Taggar
Institute of Microbial Technology
Council of Scientific and Industrial
    Research
Chandigarh, India

Mani Vasagan
Department of Marine Bio Convergence
    Science
Pukyong National University
Busan, South Korea

Bioactive Compounds and Marine Microorganisms

CHAPTER 1

Bioactive Marine Natural Products

Insights into Marine Microbes, Seaweeds, and Marine Sponges as Potential Sources of Drug Discovery

*Joseph Selvin, A. S. Ninawe, G. Seghal Kiran,
S. Arya, and S. Priyadharshini*

CONTENTS

1.1 Introduction   3
1.2 Marine Bacteria and Fungi   5
1.3 Marine Actinobacteria   6
    1.3.1   Marine Algae   6
        1.3.1.1 Potent Producers of Secondary Metabolites—Marine Algae   7
    1.3.2   Seaweed   7
    1.3.3   Marine Sponges   8
    1.3.4   Antitumor Action of Actinomycetes   9
    1.3.5   Antibacterial Action   9
    1.3.6   Antifungal Compounds   10
    1.3.7   Antiviral Compounds   10
    1.3.8   Cytotoxic Activity   10
    1.3.9   Cytostatic Activity   11
    1.3.10 Anti-Inflammatory Activity   11
    1.3.11 Antimalarial Activity   11
    1.3.12 Antidiabetic Activity   11
    1.3.13 Anti-HIV Activity   12
1.4 Conclusion and Future Perspectives   12
References   13

1.1 INTRODUCTION

Marine organisms are involved in a variety of bioactive metabolites. They manifest a great variety of biological activity. The increasing desire for medicine to be able to manage new diseases that are resistant to strains of microorganisms seemed to arouse unconventional new sources of bioactive natural products in the early 1960s. The ocean aimed outward to be appealing field. Since then numerous efforts have been made worldwide to isolate recent metabolites from microorganisms. It has been observed that marine organisms show antimicrobial, antifungal, antifertility, antiviral, antibiotic, and anticarcinogenic activities. It has also been noted that marine tunicates show a high order of antitumor, antiviral, and immunosuppressive activities. The essentialness of terrestrial microscopic organisms and parasites as wellspring of profitable bioactive metabolites has existed for a really long time. Marine Actinobacteria are the most economically and biotechnologically invaluable prokaryotes. The physiological and metabolic limits allow marine organisms to synthesize bioactive secondary metabolites in great conditions, which provides an enormous potential for the creation of novel exacerbates that are not present in living beings. The unique chemical ecology of secondary metabolite synthesis in marine organisms makes them a potential source for bioactive drugs. The significance of terrestrial microorganisms and parasites as wellsprings of significant bioactive metabolites has been exceptionally entrenched for more than a large portion of a century. Therefore, more than 120 of the most imperative medications being used today (penicillin, cyclosporin A, adriamycin, etc.) are obtained from terrestrial microorganisms. At first sight along these lines, the tremendous expected biodiversity of the strength of marine microorganisms has been the reason for the enthusiasm for their study. An extra conceivable clarification should be that marine microorganisms have constituted definitive "inviolated" outskirts for the hunt of marine natural products. In any case, albeit substantial, these were not the initial genuine reasons. Resistant sickness is the most obvious reason for death in tropical nations, contributing to about half of all fatalities. Moreover, resistant ailments and mortality rates are on the rise in developed nations (Pinner et al., 1996). Developing and redeveloping diseases are thought to be driven to a great extent by financial, natural, and environmental components (Daszak et al., 2000; Woolhouse, 2008). Between 1940 and 2004, 335 resistant ailments were reported to have been developed worldwide. These negative well-being patterns require a recharged enthusiasm for resistant sickness and also compelling methods for treatment and anticipation (Morens et al., 2004; Jones et al., 2008). Considering the emerging and reemerging infectious diseases and issues relative to antibiotic resistance, marine organisms are a rich source of novel bioactive leads for the discovery of next generation antibiotics.

The marine environment covers 70% of earth's surface and dominates the living, area accounting for 98% of potentially habitable space. Marine organisms provide a unique set of goods and services to the mankind, including climate moderation, waste and toxicant processing, and provision of vital food and medicines and other significant bioactive compounds. Bioactive compounds of marine origin have always attracted the attention of researchers and scientists globally over the last many decades. The process of bioactive molecule production by marine organisms is fascinating. Isolation of novel metabolites from marine origin leads to new innovations of a number of compounds having antimicrobial, antitumorous, antidiabetic, anti-obesity, immunological, and therapeutic potentials. Without doubt, humans have learned to satisfy their needs, to stay alive. Identification and extraction of secondary metabolites are important in two ways (Swathi et al., 2013). These microbes form highly specific and symbiotic relationships with filter feeding organisms like alcyonaria, marine plants, marine sponges, and ascidians. Since the marine environment provides a suitable biosynthetic ambient atmosphere to organisms surviving in their niches, more efficient producers can be isolated and cultured for further studies. Marine microbes play a vital role in the ecological system: they are the primary producers in the ocean, they influence the climate, and they feed on marine energy and nutrients, thus providing us with the sources of natural products and medicines.

1.2 MARINE BACTERIA AND FUNGI

Bioactive compounds, especially those derived from marine bacteria and fungi, are potential sources of novel antibiotic leads and bioactive compounds. Since the antibiotic penicillin was discovered from the fungus *Penicillium notatum*, several genera of fungi have been extensively screened for the presence of bioactive compounds (Hawksworth, 1991). Aquatic organisms have greater diversity when compared to terrestrial microbes. Marine bacteria and fungi live in close association with soft-bodied marine organisms such as sponges, thus producing bioactive secondary metabolites for chemical defense either by associated microflora or by themselves to survive in their extreme environmental conditions (Jensen and Fenical, 1991). The number of bioactive compounds discovered from marine microflora has been consistently increasing over the past few decades (Blunt et al., 2003). New marine microbiological research is thus focusing on the deep oceans and geothermal vents, a one-of-a-kind source in the world's oceans.

Marine bacteria produce secondary metabolites that thrive in harsh oceanic climates as part of bacterial antagonism. It is nature's own counteractivity for existence and survival, and these secondary metabolites serve as a wellspring of bioactive compounds for utilization in treatment of many diseases. For example, quinolone, phthalate, phenazine, phloroglucinol, pyrroles, pseudopeptide, pyrrolidinedione, phenanthren, andrimid, moiramides, bushrin, and zafrin are bioactive compounds derived from *Pseudomonas* genus of marine origin (Romanenko et al., 2008). Exopolysaccharides (EPS) from marine bacteria are another class of principal secondary metabolite having the specific properties of thickening, coagulation, adhesion, stabilization, and gelling. All these properties make them suitable for industrial applications (Satpute et al., 2007). These bacteria obtain energy from dissolved marine organic matter through cell membrane by osmosis. Bacteria are involved not only in the synthesis of secondary metabolites for defense action but also cleaning of their surroundings. A number of marine bacteria are known to produce biosurfactants and bioemulsifiers by utilizing pollutants obtained from surroundings. A marine *Bacillus circulans* strain produces a biosurfactant by utilizing anthracene, an organic pollutant, as carbon source (Das et al., 2008).

Marine microscopic organisms have regularly been considered to deliver anticancer and antibacterial compounds, permitting the biological solidness of different marine environments and the relationships between epiphytic microorganisms' surroundings governing adverse life forms and disease-causing organisms. The sharing or deleterious components that are known between these microorganisms are different, along with antimicrobial generation, lysosomes, bacteriocins, proteases, siderophores, and even pH modification through the creation of natural acids (Avendaño-Herrera et al., 2005). Explorations of Todos Santos Bay (British Columbia) revealed microscopic organisms related to ocean productivity and it was found that microbes of the families Firmicutes, Actinobacteria, and Proteobacteria produce bioactive compounds suitable for restricting the development of HCT-116 colorectal tumor cells (Villarreal-Gómez et al., 2010). Particularly, the *Bacillus* strains produce bioactive compounds with anticancer activity. Though this bacteria can develop in any substrate, it is conceivable to recommend that this species appears to have developed the ability to make compounds that hinder HCT-116 colorectal tumor cells (Aline et al., 2008).

In general, marine fungi are the least exploited resource among microorganisms, although they are a potential source of novel anti-infective agents, enzymes, biosurfactants, and other compounds having biotechnological applications (Avendaño-Herrera et al., 2005).

Marine microorganisms in general are the richest source of bioactive compounds and derivatives include analogues, non-ribosomal peptides, and hybrid molecules (Villarreal-Gómez et al., 2010). When compared with terrestrial fungi, they have developed specific secondary metabolic pathways because of their explicit living conditions, temperature variation, salinity, higher pressure, competition with other viruses, bacteria, and alternative fungi (Kossuga et al., 2012). From marine fungi, a number of fascinating compounds such as cytoglobosins and halovirs have been isolated. Thus, we have a tendency to think that there are various marine fungi containing additional exceptional structures similar to bioactive compounds (Kohlmeyer, 1984). Cyclodepsipeptide, derived from the marine fungus *Clonostachys* species, has anticancer properties (Samuel et al., 2011). About 321 types of marine fungal taxa were demonstrated until 1991, among these, 11 were under the class *Ascomycetes*, mostly found in inter-tidal and shallow waters. About 56 species of facultative fungi were described (Aline et al., 2008). Whereas, the number of marine fungal-derived bioactives were continuously increasing, it was evidence of the bioactive potential in marine fungi (Katia et al., 2012).

1.3 MARINE ACTINOBACTERIA

Marine Actinobacteria are one of the coherent groups of secondary metabolite producers and play a key role in industrial aspects. Bioprospecting marine Actinobacteria is now at the forefront in the discovery of novel antibiotic leads, since the existing antibiotics are not effective against multi-drug-resistant pathogens (Payne et al., 2006; Talbot et al., 2006). Marine Actinobacteria are known to have the ability to produce a wide variety of secondary metabolites. Actinobacteria are a unique group of secondary metabolite producers among eubacteria, and their genetic makeup of biosynthetic pathways have revealed a series of secondary metabolites (Bentley et al., 2002; Lam, 2006). About 23,000 antibiotics have been isolated from microorganisms, out of which 10,000 have been isolated from Actinobacteria. In biological sources such as mollusks, fish, sponges, seaweed, and mangroves, in addition to seawater and sediments, marine Actinobacteria are widely distributed. Marine Actinobacteria have been found to be important and prospective wellsprings of bioactive mixes, and previous studies have demonstrated that these organisms are the wealthiest wellsprings of optional metabolites. They hold a conspicuous position as they are the focus of screening programs because of their assorted qualities and their demonstrated capacity to produce novel metabolites and different molecules of pharmaceutical significance.

1.3.1 Marine Algae

Marine organisms are possible productive wellsprings of exceptionally bioactive auxiliary metabolites that may lead to helpful improvements of new pharmaceutical specialisms. Marine algae growth is not just utilized as a part of a discovery for new medications; they are likewise utilized widely as nourishment in different spots. Algae can be arranged into two principal bunches, which include the first group—blue-green microalgae and diatoms, the second group encompasses green and brown microalgae. The microalgae are a potential source of bioactive pigments and biofuel. Their bioactive pigments and the whole biomass as a single cell protein source were developed as marine-derived nutraceuticals. Microalgal-derived bioactive compounds and biomass can be developed as tonic foods and nutritional supplements. From a monetary perspective, marine algae growth is an essential asset for nourishment and mechanical information. The seashore consists of diverse species that are used as nourishment for humans, therapeutic items, manure, and fuel, and assume a critical part in the extraction of phycocolloids and hydrocolloids (Teas, 2007). One of these has a noteworthy mechanical application, despite the theory on the sea development potential as a prompt wellspring of proteins and pharmaceutical items.

1.3.1.1 Potent Producers of Secondary Metabolites—Marine Algae

Marine algae are an appetizer of halogenated metabolites with potential business esteem. Structures displayed by these compounds run from noncyclic elements with a straight fasten to complex polycyclic particles (Salvador et al., 2007; Kandhasamy and Arunachalam, 2008; Manilal et al., 2009). Their restorative and pharmaceutical applications have been examined for a couple of decades; however, different properties, for example, antifouling, are not to be disposed of. Numerous compounds were found in the most recent years, in spite of the fact that the requirement for new medications keeps this field open. The same number of algal species are ineffectively screened. The biological part of marine algal halogenated metabolites has been ignored some way or the other. This new research field will give a profitable and novel understanding of the marine biological community flow and another way to deal with fathoming biodiversity. Also, understanding the cooperation between auxiliary compound creation by algae growth and the earth, including anthropogenic or worldwide atmosphere changes, is a crucial focus in the coming years (Arasaki and Arasaki, 1983; John Davis and Vasanthi, 2011). Examination of secondary metabolites has been more centered on macroalgae than on phytoplankton. Notwithstanding, phytoplankton could be an extremely encouraging material since it is the marine's base evolved way of life with brisk adjustment to natural changes, which without a doubt has results on optional digestion system. This chapter surveys recent developments in this field and introduces patterns of marine algae as producers of secondary metabolites.

1.3.2 Seaweed

Seaweeds are plant varieties that are found in the seas and oceans. They are multicellular, macroscopic algae that are present near the seabed. They are a rich source of polysaccharide, minerals, and certain vitamins (Arasaki and Arasaki, 1983). Bioactive compounds present in seaweed include polysaccharides, proteins, lipids, and polyphenols, with antibacterial, antiviral, antifungal, and other properties (Kumar et al., 2008b). This makes seaweed a potent supplement in functional foods or for the extraction of compounds.

Marine algae contain an appreciable amount of polysaccharides, including cell wall components and storage polysaccharides (Murata and Nakazoe, 2001; Kumar et al., 2008a). The commercial applications of these can be as stabilizers, thickeners, emulsifiers, food, feed, beverages, and so on. The aggregate polysaccharide fixations in seaweed is between 5% and 75% of dry weight. The highest polysaccharide content was found in green alga *Ulva*, *Ascophyllum*, and *Porphyra*. Increased health consciousness has been one of the most important stimulating factors for rapid growth of the functional food industry. Herein, seaweed sources have received much attention since seaweeds are a rich source of nutritional supplements, antioxidants, and bioactives. Notably, fucoidans have shown evidence of playing a vital role in human health and nutrition due to their numerous biological activities and health benefit effects. Thus, extensive studies of fucoidans will discover novel biological properties as well as novel functional applications in pharmaceuticals, nutraceuticals, and functional foods. Fucoidans are a complex series of sulfated polysaccharides found widely in the cell walls of brown seaweeds. Such polysaccharides do not occur in other divisions of algae or in terrestrial plants. Fucoidans have been found to possess various functional activities including antioxidant, anti-inflammatory, anti-allergic, antitumor, anti-obesity, and anticoagulant effects (Murata and Nakazoe, 2001; Kumar et al., 2008a). Algal-derived bioactive compounds including sulfated polysaccharides, fucoidans, mannitol, and so forth, vary along with the seasons and geographical locations. The seasonality and variations in secondary metabolite contents of seaweeds has been well documented (Selvin, 2002; Selvin, 2010; Kraan, 2012).

Seaweed have about 32%–66% of dietary fibers on dry weight basis which are rich in soluble fraction (Lahaye, 1991). They are of two types: water-insoluble compounds like xylans and mannans and water-soluble dietary fibers like agar, alginic acid, and porphyran. Seaweed dietary fibers show antitumor and antiherpetic activity. They are potent anticoagulants and also reduce LDL cholesterol levels. They anticipate digestive organ disease and diabetes and furthermore have antiviral properties (Murata and Nakazoe, 2001; Lee et al., 2004). Protein content in seaweed shifts among species. Higher protein substances are found in green and red green growth, and lower protein substances are found in brown algae growth. *Undaria* has the most extreme substance with 24% of dry weight, while *Ascophyllum* has the least with 10% of dry weight. Normal protein content changes somewhere around 17% and 21%. The protein content in a few ocean growths are considerably higher than profoundly protein-aceous vegetables such as soybeans, which have 40% protein content, while it is up to 44% in rich species. They are a rich wellspring of key amino acids and acidic amino acids like aspartic corrosive and glutamic corrosive (Fleurence, 2004). In spite of the fact that threonine, lysine, tryptophan, sulfur-containing amino acids, and histidine are restricting amino acids in star teins, their level in algal proteins is higher than in terrestial plants (Galland-Irmouli et al., 1999).

The unsaponifiable part of seaweeds contains unsaturated fats and carotenoids like β-carotene, lutein, violaxanthin in red and green growth, and fucoxanthin in brown algae. They likewise contain tocopherols and sterols (Jensen, 1969). Lipid content in seaweed is around 4.5%, which is lower than other marine sources and is consequently a low nourishment vitality source (Murata and Nakazoe, 2001). Green growth can gather polyunsaturated unsaturated fats (PUFAs) amid reduction in ecological temperature (Khotimchenko, 1991). Omega-3 unsaturated fat generation has additionally been accounted for in ocean growth. The prevalent unsaturated fat in different seaweed products is EPA (C20:5, *n*-3), which focuses as high as half of the aggregate unsaturated fat substance (Murata and Nakazoe, 2001). The measure of phospholipids in red algae changes from 10% to 20% of aggregate lipid content. The real phospholipid is phosphatidylcholine (62%–78%). The red algae contains practically meet amounts of mono-glycosyldiacylglycerol (MGDG), diglycosyldiacylglycerol (DGDG), and sulfaquinovosyl-diacyl-glycerol, with a few special cases. MGDG and DGDG are the major glycolipids in green seaweed.

1.3.3 Marine Sponges

Marine secondary metabolites are the organic compounds that are produced by microbes, seaweed, and some other marine organisms (Attaway and Zaborsky, 1993). The host microorganism synthesize these substances as non-primary or secondary metabolites, and it can maintain homeostasis in the environment (Selvin, 2002). Secondary metabolites isolated from marine samples can possess unique structure, and they have a wide range of biological activities. Among marine organisms, sponges are the richest phyla in taxonomic species. Sponges produce metabolites to repulse and dissuade predators, and contend with different species. Some remedial aggravates that have been distinguished in sponges incorporate anticancer operators and immunomodulators. Some sponges can likewise create possibly helpful antifouling operators. Bioactive secondary metabolites segregated from sponges deliver a practical compound group, which can be from sponges and their microorganisms. Sponges produce chemicals by transmitting bodily fluid containing poisons to repel other attacking predators (Baker et al., 2007).

Marine sponges belong to the genus *Ircinia*, which is a rich source of terpenoids and several have shown a wide variety of biological activity. Terpenoids contain a tetronic acid moiety, which has strong antibiotic activity. Components of marine sponges are known to modulate numerous biological activities and possess anticancer, anti-inflammatory, and antifungal effects. Among halogenated alkaloids, bromoalkaloids form a widely used group of natural compounds that can be predominantly seen in marine eukaryotes like sponge. Sponges are a rich source of natural products or secondary metabolites, which can be used for pharmaceutical and biotechnological applications.

1.3.4 Antitumor Action of Actinomycetes

Malignancy is a disease that includes more than a hundred sorts. The frequency of malignancy expands continually, posing a serious threat to the well-being of organizations. In a large number of cases, the solutions utilized as a part of chemotherapy treatment incite optional danger or resistance. Anticancer activity of actinomycete metabolites now and actinomycin D earlier were one of the primary trademark metabolites used to treat tumors. Other anticancer agents, such as anthracyclines, daunorubicin (daunomycin), and doxorubicin (Adriamycin), were acquired in the 1960s in the midst of the biosynthesis strategy of *S. peucetius*. Epirubicin is one of the more startling recognized anthracycline gathering blends; it was affirmed by the FDA in 1999 and has a more predominant therapeutic profile than doxorubicin in light of its less-unfavorable effects. It is used as a piece of treatment of chest development, ovarian harm, lung ailment, and leukemia. Bleomycin is another antitumor compound. Mitomycins indicate high antitumor development. On the other hand, their usage is compelled in treatment on account of their toxic quality. Streptozotocin from *S. achromogenes*, indicates particular harm against pancreatic β cells. It was affirmed by the FDA in 1982 as a pancreatic islet cell antitumor drug. *Calicheamicins* are antitumor blends confined from *Micromonospora echinospora*. They act by isolating DNA in its twofold-stranded condition. They are exceedingly noxious; however, in the occasion that associated to immunizer that particularly passes on pharmaceuticals to tumor cells, they have shown a deadly antitumor effect.

1.3.5 Antibacterial Action

Marine Actinobacteria are well known for their antibacterial activity. Marine *Verrucosispora* strain has produced a novel polycyclic polyketide antibiotic, Abyssomicin C, which intensively acts against Gram-positive microscopic organisms, including clinical confines of multiple resistant and vancomycin-resistant *Staphylococcus aureus*.

Marine Actinobacteria serve as a resource of numerous novel antibacterial drugs such as abyssomicins, bonactin, chloro-dihydroquinones, diazepinomicin, frigocyclinone, essramycin, lynamicins, marinopyrroles, caboxamycin, himalomycins, marinomycins, glyciapyrroles, tirandamycin, bisanthraquinone, gutingimycin, helquinoline, lajollamycin, lincomycin, tirandamycins, and 1,4-dihydroxy-2-(3-hydroxybutyl)-9,10-anthraquinone.

1.3.6 Antifungal Compounds

Amphotericin B, a champion among the most known actinomycete metabolites, is a macintosh rolide polyene antimicrobial that was confined in 1955 from *S. nodosus*. It is an antifungal anti-infection with an extensive variety of development, nevertheless, its various indications limit its clinical application. Two classes of antifungal nucleosides seriously stifle infectious chitin synthase: polyoxins and nikkomycins. Nikkomycins (nikkomycin Z) are more intense against *Candida albicans* than polyoxins. Nikkomycins are gotten from *S. tendae* and *S. ansochromogenes*. Nikkomycins were withdrawn in the 1970s from *S. tendae* and ended up being against *Rhizopus circinans* and additionally against *Botrytis cinerea*. Nikkomycins X and Z have a spot to peptidyl nucleoside immunizing agents and are gotten from *S. ansochromogenes*. Nikkomycins go about as inhibitors of chitin synthase.

1.3.7 Antiviral Compounds

From 1998 to 2008, an enormous amount of marine life structures were considered for the isolation of antiviral compounds. Numerous species demonstrated striking activities against pathogenic viral infections. In any case, this was seen in vitro. Chemical grouping of marine compounds was done to characterize them into the following real marine substance mixes: polyketides, terpenes, nitrogen-containing compounds, and polysaccharides. From the sponge *Lissodendoryx* sp. antiviral compounds, halichondrin B, homohalichondrin B, and isohomohalicondrin B were derived. In vivo and in vitro trials have been held for inhibition of the proliferation of tumor cells after replication by these compounds (Munro et al., 1987). At present, it shows up as though only several compounds have been acquired from marine microorganisms in light of the antiviral developments, remembering that cyanobacteria are accumulated as blue–green algae growth. Contemplates using marine bacteria as hotspots for antiviral compounds have focused for the most part on marine bacterial exopolysaccharides (Arena et al., 2006).

1.3.8 Cytotoxic Activity

Four new polyketides, salinipyrones A and B and pacificanones A and B, have been isolated from marine Actinobacteria *Salinispora pacifica* CNS-237, found in the buildup assembled from Palau Island in the Western Pacific Ocean. Natural activities of these compounds are being examined in various bioassays. In the underlying screening, salinipyrones and pacificanones demonstrated no huge development in an illness cytotoxicity test using HCT-116 human colon tumor cells. In a disengaged mouse splenocyte model of unfavorably powerless bothering, salinipyrone A indicated moderate restriction of interleukin-5 creation significantly at 10 g/mL without quantifiable human cell cytotoxicity.

1.3.9 Cytostatic Activity

Proximicins are aminofuran anti-infection agents, no doubt arranged by NRPS outline work, made by *Verrucosispora* strain. The trademark essential segment of proximicins is 4-amino-furan-2-carboxylic acid, up to this point dark c-amino acid (Schneider et al., 2008). Proximicins A, B, and C have cytostatic impacts against different tumor cell lines. Proximicins A, B, and C indicated a huge development movement against human gastric adenocarcinoma AGS (GI50 of 0.6, 1.5, and 0.25 M, individually) and hepatocellular carcinoma Hep G2 (GI50 of 0.82, 9.5, and 0.78 M, separately) (Fiedler et al., 2008; Schneider et al., 2008), and were found to affect the capture of AGS cells in G0/G1 and to expand the levels of p53 and p21.

1.3.10 Anti-Inflammatory Activity

Cyclomarin A is another cyclic heptapeptide anti-infection secluded from *Streptomyces* sp. It showed critical calming action in both in vivo and in vitro examinations (Renner et al., 1999). Salinamides A and B are bicyclic depsipeptides, derived from *Streptomyces* sp. CNB-091, secluded from the jellyfish *Cassiopeia xamachana*. These metabolites are valuable as antimicrobial and calming specialists (Moore et al., 1999).

1.3.11 Antimalarial Activity

Trioxacarcins are unpredictable mixes indicating higher antimalarial action against malarial pathogens, and some of them have high antitumor and antibacterial activity. Trioxacarcin A, B, and C were obtained from *Streptomyces ochraceus* and also *Streptomyces bottropensis* (Maskey et al., 2004). Some of these compounds have amazingly high antiplasmodial action, which is similar to that of artemisinin, the most dynamic compound against pathogens causing intestinal sickness.

1.3.12 Antidiabetic Activity

Though the marine world has attracted the attention of the research community for years, the therapeutic potentiality of marine phytoplanktons is basically undiscovered. Enzymes play a morbific role as underlying explanation for many diseases like polygenic disorders, Alzheimer, and Parkinson (Bhattacharjee et al., 2014). Enzymes like alpha-glycosidase, a simple sugar enzyme, have been found to cause pathogenicity of two polygenic disorders and their associated complications. Therapeutically active agents like bromophenols, 2-piperidione, benzeneacetamide, *n*-hexadecanoic acid, fucoxanthin, dysidine, biologically active internal secretion, methyl-ethyl organic compound derivatives, and so on, were explored from marine species. Astonishing resources of marine bioactivities are from species like sponges (31%), red algae (4%), brown algae (5%), green algae (1%), microorganisms (15%), coral (24%), ascidians (6%), Mollusca (6%), and others (8%). Many marine red and green algae are found to be potential inhibitors of α-glucosidase, aldose reductase (AR), and protein tyrosine phosphatase (PTP). Bromophenols found in some red algae like *Rhodomela confervoides*, *Symphyocladia latiuscula*, and *Polysiphonia urceolata* show significant hypoglycemic potentials by inhibiting PTP, α-glucosidase, and AR apart from their antioxidant activity.

1.3.13 Anti-HIV Activity

Human immunodeficiency virus (HIV) reasons obtained resistant inadequacy disorder (AIDS) and is a worldwide general well-being issue. Hostile to HIV treatment including synthetic medications have enhanced the life nature of HIV/AIDS patients [53]. On the other hand, development of HIV drug resistance, reactions, and the need for long haul hostile HIV treatment is the fundamental purpose behind the disappointment of hostile HIV treatment. Along these lines, it is fundamental to segregate novel hostile HIV therapeutics from normal assets. A great deal of time has been spent on marine-decided threatening HIV masters, for instance, phlorotannins, sulfated chitooligosaccharides, sulfated polysaccharides, lectins, and bioactive peptides. This dedication acquaints a survey of antagonistic HIV therapeutics from marine resources and their potential application in HIV treatment. Phlorotannins, which have been found to exist within chestnut green growth development, are encircled by the polymerization of phloroglucinol (1,3,5-tryhydroxybenzene) monomer units and biosynthesized through the acetatemalonate pathway. The phlorotannins are incredibly hydrophilic parts with a broad assortment of sub-nuclear sizes stretching between 126 Da and 650 kDa. Marine brown algae growth results in an accumulation of phloroglucinol-based polyphenols, as phlorotannins of low, center and high nuclear weight containing both phenyl and phenoxy unit. In addition, phlorotannins contain phloroglucinol units associated with each other in various ways, and are of wide occasion among marine life forms, particularly chestnut and red green growth development. Considering the strategy for linkage, phlorotannins can be divided into four subclasses: fuhalols and phlorethols (phlorotannins with an ether linkage), fucols (with a phenyllinkage), fucophloroethols (with an ether and phenyl linkage), and eckols (with adibenzodioxinlinkage). Cocoa green growth development, which has a high gathering of phlorotannin, have been accounted for to be against HIV movement.

1.4 CONCLUSION AND FUTURE PERSPECTIVES

Characteristic products are an essential asset for the elaboration of pharmaceuticals. Despite the way that a paramount number of bioactive compounds, life structures, and marine assets have been assessed in the journey for new bioactive compounds, it has been insufficient and it is key and fundamental to keep on searching for new metabolites, particularly those that are endophyte microorganisms of ocean improvement. The unpleasant usage given to hostile to contamination specialists has realized the change of tiny life forms strains that are impenetrable to a huge bit of the known meds. This situation has led to an obliged mission for new against microbial mixes, being the seabed a confident site for examination and future medicine headway. Moreover, the chemotherapy treatment for the varying explanations behind differing sorts of ailments that are present today, appears compelling so the examination becomes vital in the regular's science items. The routines for bacterial society and recognizable proof have turned out to be exceptionally encouraging particularly, those done through atomic strategies, by which it is conceivable to distinguish a strain up to animal types and once in a while at a subspecies level. The assorted connections that exist in the middle of microorganisms and their visitors incite that bacterial mixes can in the end be utilized as a wellspring of new medications for human prosperity. Significant increments in the enthusiasm relating to natural products, particularly those starting from marine organisms, and demonstrated capability of marine organisms to orchestrate differing compounds have brought marine organisms into the core interest of concentrated examination. It appears to be likely that two noteworthy parallel methodologies will be created for medication disclosure from marine organisms. One will be based on the foundation of new and further improvement of existing separation and development systems, to build the differing qualities of cultivable segregates, abbreviate the season of development to accomplish calculable cell masses for higher yields, and the creation of auxiliary metabolites. As the marine sources vary exceptionally and every species may have uncommon prerequisites not just for development, additionally for the generation of auxiliary metabolites, this methodology will require significant novel and inventive endeavors. Its real leverage is that, if fruitful, it will bring about a generation of an intensified study whose structure, curiosity, and organic action can be surveyed immediately. As the improvement subordinate development is implausible for revealing the full biosynthetic potential, the second approach, genome-based bioprospecting, must be held onto in order to address this request. Its prosperity will rely upon variables, for example, the advancement of productive instruments for bioinformatic examination of the genomes permitting distinguishing proof of interesting biosynthetic quality groups, chemoinformatic programming for the expectation of auxiliary elements taking into account quality investigation and further advancement of host/vector frameworks for heterologous articulation of quality bunches. A solid transaction of established normal item science with cutting-edge microbial hereditary qualities and bioinformatics will, without a doubt, help defeat the supply and manageability issues of the past and help to advance the generation of bioactive substances from marine living beings as a very much perceived elective for future medication revelation programs.

REFERENCES

Aline, M.V.M., Simone, P.L., and Roberto, G.S.B. A multi-screening approach for marine derived fungal metabolites and the isolation of cyclodepsipeptides from *Beauveria feline*. *Quim. Nova.*, 31(5) (2008): 1099–1103.

Arasaki, S. and Arasaki, T. *Low Calorie, High Nutrition Vegetables from the Sea to Help You Look and Feel Better*. Japan Publications, Tokyo, Japan, 1983, 196 pp.

Arena, A., Maugeri, T.L., Pavone, B., Iannello, D., Gugliandolo, C., and Bisignano, G. Antiviral and immunomodulatory effect of a novel exopolysaccharide from a marine thermotolerant *Bacillus licheniformis*. *Int. Immunopharmacol.*, 6 (2006): 8–13.

Attaway, D.H. and Zaborsky, O.R. *Marine Biotechnology*, Vol. I: *Pharmaceutical and Bioactive Natural Products*. Plenum Press, New York, 1993, pp. 1–44.

Avendaño-Herrera, R., Lody, M., and Riquelme, C.E. Producción de substancias inhibitorias entre bacterias de biopelículas en substratos marinos. *Revista Biología Marina y Oceanografía*, 40(2) (2005): 117–125.

Baker, D.D., Chu, M., Oza, U., and Rajgarhia, V. The value of natural products to future pharmaceutical discovery. *Nat. Prod. Rep.*, 24 (2007): 1225–1244.

Bentley, S., Chater, K., Cerdeno-Tarraga, A.-M. et al. Complete genome sequence of the model actinomycete *Streptomyces coelicolor* A3 (2). *Nature*, 417(6885) (2002): 141–147.

Bhattacharjee, R., Mitra, A., Dey, B., and Pal, A. Exploration of anti-diabetic potentials amongst marine species—A mini review. *Indo Global J. Pharm. Sci.*, 4(2) (2014): 65–73.

Blunt, J.W., Copp, B.R., Munro, M.H.G., Northcote, P.T., and Prinsep, M.R. Marine natural products. *Nat. Prod. Rep.*, 20 (2003): 1–48.

Das, P., Mukherjee, S., and Sen, R. Improved bioavailability and biodegradation of a model polyaromatic hydrocarbon by a biosurfactant producing bacterium of marine origin. *Chemosphere*, 72 (2008): 1229–1234.

Daszak, P., Cunningham, A.A., and Hyatt, A.D. Emerging infectious diseases of wild-life—Threats to biodiversity and human health. *Science*, 287 (2000): 443–449.

Fiedler, H.-P., Bruntner, C., Riedlinger, J. et al. Proximicin A, B and C, novel aminofuran antibiotic and anticancer compounds isolated from marine strains of the actinomycete *Verrucosispora*. *J. Antibiot.*, 61(3) (2008): 158–163.

Fleurence, J. Seaweed proteins. In: Yada, R.Y. (ed.), *Proteins in Food Processing*. Woodhead Publishing, Cambridge, UK, 2004, pp. 197–213.

Galland-Irmouli, A.V., Fleurence, J., Lamghari, R., Luçon, M., Rouxel, C., Barbaroux, O., Bronowicki, J.P., Villaume, C., and Guéant, J.L. Nutritional value of proteins from edible seaweed *Palmaria palmata* (dulse). *J. Nutr. Biochem.*, 10 (1999): 353–359.

Hawksworth, D.L. The fungal dimension of biodiversity: Magnitude, significance and conservation. *Mycol. Res.*, 95 (1991): 641–655.

Jensen, A. Tocopherol content of seaweed and seaweed meal. 3. Influence of processing and storage on content of tocopherols, carotenoids and ascorbic acid in seawood meal. *J. Sci. Food Agric.*, 20 (1969): 622–626.

Jensen, P.R. and Fenical, W. Strategies for the discovery of secondary metabolites from marine bacteria: Ecological perspectives. *Annu. Rev. Microbiol.*, 48 (1991): 559–584.

John Davis, G.D. and Vasanthi, A.H.R. Seaweed metabolite database (SWMD): A database of natural compounds from marine algae. *Bioinformation*, 5(8) (2011): 361–364.

Jones, K.E., Patel, N.G., Levy, M.A. et al. Pharmaceutically active secondary metabolites of marine actinobacteria Global trends in emerging infectious diseases. *Nature*, 451(7181) (2008): 990–993.

Kandhasamy, M. and Arunachalam, K.D. Evaluation of in vitro antibacterial property of seaweeds of southeast coast of India. *Afr. J. Biotechnol.*, 7(12) (2008): 1958–1961.

Katia, D., Teresa, A.P.R.S., Ana, C.F., and Armando, C.D. Analytical techniques for discovery of bioactive compounds from marine fungi. *Trends Anal. Chem.*, 34 (2012): 97–110.

Khotimchenko, S.V. Fatty-acid composition of 7 *Sargassum* species. *Phytochemistry*, 30 (1991): 2639–2641.

Kohlmeyer, J. Tropical marine fungi. *Mar. Ecol.*, 5 (1984): 329–378.

Kossuga, M.H., Romminger, S., Xavier, C. et al. Evaluating methods for the isolation of marine-derived fungal strains and production of bioactive secondary metabolites. *Revista Brasileira da Farmacognosia Braz. J. Pharmacogn.*, 22(2) (2012): 257–267.

Kraan, S. Algal polysaccharides, novel applications and outlook. In: Chang, C.F. (ed.), *Carbohydrates-comprehensive studies on glycobiology and glycotechnology*, InTech, New York, 2012.

Kumar, C.S., Ganesan, P., and Bhaskar, N. In vitro antioxidant activities of three selected brown seaweeds of India. *Food Chem.*, 107 (2008a): 707–713.

Kumar, C.S., Ganesan, P., Suresh, P.V., and Bhaskar, N. Seaweeds as a source of nutritionally beneficial compounds—A review. *J. Food Sci. Technol.*, 45 (2008b): 1–13.

Lahaye, M. Marine-algae as sources of fibers—Determination of soluble and insoluble dietary fiber contents in some sea vegetables. *J. Sci. Food Agric.*, 54 (1991): 587–594.

Lam, K.S. Discovery of novel metabolites from marine actinomycetes. *Curr. Opin. Microbiol.*, 9(3) (2006): 245–251.

Lee, J.B., Hayashi, K., Hashimoto, M., Nakano, T., and Hayashi, T. Novel antiviral fucoidan from sporophyll of *Undaria pinnatifida* (Mekabu). *Chem. Pharm. Bull.*, 52 (2004): 1091–1094.

Manilal, S., Sujith, S., SeghalKiran, G., Selvin, J., Shakir, C., Gandhimathi, R., and Nataraja Panikkar, M.V. Biopotentials of seaweeds collected from southwest coast of India. *J. Mar. Sci. Technol.*, 17(1) (2009): 67–73.

Maskey, R.P., Helmke, E., Kayser, O. et al. Anti-cancer and antibacterial trioxacarcins with high anti-malaria activity from a marine Streptomycete and their absolute stereochemistry. *J. Antibiot.*, 57(12) (2004): 771–779.

Moore, B.S., Trischman, J.A., Seng, D., Kho, D., Jensen, P.R., and Fenical, W. Salinamides, antiinflammatory depsipeptides from a marine streptomycete. *J. Org. Chem.*, 64(4) (1999): 1145–1150.

Morens, D.M., Folkers, G.K., and Fauci, A.S. The challenge of emerging and re-emerging infectious diseases. *Nature*, 430(6996) (2004): 242–249.

Munro, M.H.G., Luibrand, R.T., and Blunt, J.W. The search for antiviral and anticancer compounds from marine organisms. In: Scheuer, P.J. (ed.), *Bioorganic Marine Chemistry*, Vol. 1. Verlag Chemie, Berlin, Germany, 1987, pp. 93–176.

Murata, M. and Nakazoe, J. Production and use of marine algae in Japan. *Jpn. Agric. Res. Q.*, 35 (2001): 281–290.

Payne, D.J., Gwynn, M.N., Holmes, D.J., and Pompliano, D.L. Drugs for bad bugs: Confronting the challenges of antibacterial discovery. *Nat. Rev. Drug Discov.*, 6(1) (2006): 29–40.

Pinner, R., Teutsch, S., Simonsen, L., Klug, L., Grabers, J., Clarke, M., and Berkelman, R. Trends in infectious diseases mortality in the United States. *J. Am. Med. Assoc.*, 275 (1996): 189–193.

Renner, M.K., Shen, Y.-C., Cheng, X.-C. et al. New antiinflammatory cyclic peptides produced by amarine bacterium (*Streptomyces* sp.). *J. Am. Chem. Soc.*, 121(49) (1999): 11273–11276.

Romanenko, L.A., Uchino, M., Tebo, B.M., Tanaka, N., Frolova, G.M., and Mikhailov, V.V. *Pseudomonas marincola* sp. nov., isolated from marine environments. *Int. J. Syst. Evol. Microbiol.*, 58 (2008): 706–710.

Salvador, N., Gómez-Garreta, A., Lavelli, L., and Ribera, L. Antimicrobial activity of Iberian macroalgae. *Sci. Mar.*, 71 (2007): 101–113.

Samuel, P., Prince, L., and Prabakaran, P. Antibacterial activity of marine derived fungi collected from south east coast of Tamilnadu. *India J. Microbiol. Biotech. Res.*, 1(4) (2011): 86–94.

Satpute, S.K., Banat, I.M., Dhakephalkar, P.K., Banpurkar, A.G., and Chopade, B.A. Biosurfactants, bioemulsifiers and exopolysaccharides from marine microorganisms. *Biotechnol. Adv.*, 28 (2007): 436–450.

Schneider, K., Keller, S., Wolter, F.E. et al. Proximicins A, B, and C—Antitumor furan analogues of netropsin from the marine actinomycete Verrucosispora induce upregulation of p53 and the cyclin kinase inhibitor p21. *Angew. Chem. Int. Ed.*, 47(17) (2008): 3258–3261.

Selvin, J. Shrimp disease management using secondary metabolites isolated from marine organisms. PhD thesis submitted to M.S. University, Tirunelveli, India, 2002, 204pp.

Selvin, J. *Development of an Integrated Disease Management (IDM) System for Sustainable Shrimp Farming.* Project completion report, Department of Biotechnology (DBT), New Delhi, 2010, pp. 90.

Swathi, J., Narendra, K., Sowjanya, K.M., and Krishna Satya, A. Marine fungal metabolites as a rich source of bioactive compounds. *Afr. J. Biochem. Res.,* 7(2013): 184–196.

Talbot, G.H., Bradley, J., Edwards, J.E., Gilbert, D., Scheld, M., and Bartlett, J.G. Bad bugs need drugs: An update on the development pipeline from the antimicrobial availability task force of the infectious diseases society of America. *Clin. Infect. Dis.,* 42(5) (2006): 657–668.

Teas, J. Seaweed and soy: Companion foods in Asian cuisine and their effects on thyroid function in American women. *J. Med. Food.,* 10 (2007): 90–100.

Villarreal-Gómez, L.J., Soria-Mercado, I.E., Guerra-Rivas, G., and Ayala-Sánchez, N.E. Antibacterial and anticancer activity of seaweeds and bacteria associated with their surface. *Revista de Biología Marina y Oceanografía,* 45(2) (2010): 267–275.

Woolhouse, M.E. Epidemiology: Emerging diseases go global. *Nature,* 451(7181) (2008): 898–899.

SECTION II

Nutraceuticals

CHAPTER 2

Isolation and Screening for Bioactive Compounds

Rama Bhadekar, Anuradha Mulik, and Sonali Ambulkar

CONTENTS

2.1   Introduction   19
2.2   Sample Collection and Isolation   20
    2.2.1   Isolation of Extremophiles   21
    2.2.2   Media   22
2.3   Screening for Industrially Important Bioactive Compounds   23
    2.3.1   Enzymes   23
    2.3.2   Enzyme Inhibitors   24
    2.3.3   Exopolysaccharides   25
    2.3.4   Biosurfactants   25
    2.3.5   Polyunsaturated Fatty Acids   25
    2.3.6   Pigments   26
    2.3.7   Marine Probiotics   26
2.4   Screening for Biomedically Important Compounds   27
    2.4.1   Antibacterial Activity   27
    2.4.2   Antifungal Activity   28
    2.4.3   Anti-HIV Activity   28
    2.4.4   Anticancer Activity   28
    2.4.5   Antioxidant Activity   29
2.5   Conclusion   29
References   30

2.1 INTRODUCTION

Today, biotechnological procedures are gaining importance in various sectors, such as pharmaceuticals, food, cosmetics, agriculture, and so on, due to their sustainability, cost-effectiveness, and eco-friendly nature. These methods are studied and optimized to achieve superior quality and quantity of the product as compared to chemical methods. Chemical synthesis strategies have certain limitations, such as environmental pollution, excessive use of chemicals, adverse effects on human health, and high cost. Therefore, plants, animals, and microorganisms are being scrutinized as natural resources of bioactive compounds and are being harvested in the last few years.

Among these, microorganisms are of tremendous value because of their fast growth rate, ubiquitous presence, and ease of optimization. Bacteria, fungi, and actinomycetes are widely explored for the range of bioactive compounds they offer that have various industrial as well as biomedical applications. These microorganisms may be sourced from various ecosystems, such as terrestrial, freshwater, and marine. They are found to adapt to the environmental conditions in their respective habitats. Of these three ecosystems, marine ecosystem is the largest since water accounts for more than 70% of the earth's surface and marine environment accounts for more than 97% of that (Foucher, 2009). The peculiarity of marine environment can be attributed to the following properties of marine water: (1) higher density and viscosity than air, (2) better ability to transmit sound, (3) low electrical resistivity, (4) ability to absorb light, (5) variation in oxygen concentration with temperature and salinity, and (6) very low oxygen diffusion rate (Nybakken and Bertne, 2004).

This chapter elaborates bioprospecting of marine microorganisms with respect to their isolation and screening for bioactive compounds. For the isolation of microorganisms, various samples from the marine environment prove to be useful sources. They include seawater at different depths, sea bottom, deep sea hydrothermal vent, microorganisms present in symbiotic association with marine plants and animals, fouling ship hulls, and so on. Usually, seawater contains 1 million microorganisms/mL. Marine microorganisms survive under conditions of high pressure, low temperature, and high salinity. It may be one of the reasons for the isolation of a majority of Gram-negative bacteria from marine environment—their cell envelope is well suited to such surroundings. As a result, bioactive compounds obtained from marine microorganisms are found to withstand harsh conditions such as extreme pH and temperature, high salt concentration, and high pressure.

2.2 SAMPLE COLLECTION AND ISOLATION

The isolation of marine microorganisms is usually aimed at the selection of better strains which have an ability to produce novel bioactive metabolites with unique structural characteristics. Common isolation procedures involve sample collection, transport to the laboratory, enrichment, and serial dilution, followed by plating on respective media. However, the method cannot be generalized as it can vary with the sample and may or may not require enrichment. Soil, marine sediments, water, sea animals, mangroves, and seaweed are starting materials reported thus far. Next, we describe methods to collect these samples.

The collection of marine sediments is carried out by an alcohol-rinsed Peterson grab followed by a transfer of samples into ziplock bags using a sterile spatula (Dhevendaran and Anithakumarai, 2002). Samples from deep sea sediments are also collected aseptically using trawls or push core samplers (da Silva et al., 2013), which are transferred to falcon tubes and stored at 4°C. The filter-sterilized seawater can be used as a diluent for serial dilution. For bacterial isolation from soil samples, soil suspension is prepared by diluting 10 g of sample in 90 mL sterile water followed by shaking to get a homogenized suspension. However, for microbial isolation from a mangrove soil sample (rhizosphere and non rhizosphere), the samples are pretreated with dry heat or chemicals such as phenol or wet heat in sterile seawater at 50°C for 15 min. The pretreatment of soil is followed by 1:10 dilution (V/V) with sterile ¼ strength Ringer's solution and serial dilution (Hong et al., 2009).

Seawater samples are normally collected from the depth of 5 to 25 m. This is again followed by serial dilution and plating. Filtered autoclaved seawater and/or synthetic seawater medium are used for subculturing. Marine nutrient agar and Zobell Marine Agar can be used for spreading and isolation (Pimpliskar and Jadhav, 2014).

For the isolation of microorganisms from mangrove ecosystems, different plant parts, such as the root, stem, leaf, fruit, and flower, are used and samples are stored at 4°C until processed. The plant tissue surface is cleaned using sterile water and air-dried in laminar air flow, followed by 70% ethanol treatment for 5 min, 0.1% $HgCl_2$ treatment for 15 min, and 5 times washing with 1% Tween® 80 each for 5 min. Then, the small pieces are cut and ground using mortar and pestle in 50% sterile seawater, and this preparation can be used for plating on selective media. The preference is to collect the samples in clean polythene bags, transport them to the laboratory, and process within 3 h (Hong et al., 2009). Artificial seawater or sterile seawater can also be used for repeated washings of mangrove leaf samples or seaweed tissue (Bonugli-Santos et al., 2015). Prior to the isolation of endophytic bacteria, the samples are kept in plastic bags, washed with water, and surface disinfected, and the cut pieces are used to prepare suspension in phosphate-buffered saline by shaking for 1 h (da Silva et al., 2013). This is followed by serial dilution and plating on tryptic soy agar. Incubation is carried out at 28°C until growth.

Surface sterilization with $HgCl_2$ or ethanol followed by washing with sterile seawater is suggested for marine invertebrate samples (Newel, 1976). However, better microbial recovery is observed if $HgCl_2$ is eliminated. Marine microorganisms have also been isolated from animals such as fish, sponges, corals, and so on. Sponge samples are collected in an ethyl polythene bag by SCUBA diving from the depth of 5 to 10 m and transferred to the laboratory ascetically in ice (Devi et al., 2010). Isolation involves surface sterilization of the sponge and removal of the sponge tissue, followed by maceration and homogenization and plating on Zobell Marine Agar after dilution. Incubation temperature is 27°C–30°C. For the isolation of microbes associated with coral reefs, the coral tissue is placed in sterile plastic bags, placed in ice, and brought to the laboratory. Further procedures involve washing with sterile seawater and the preparation of a suspension of the coral sample in sterile seawater. As stated, serial dilution and plating on artificial seawater nutrient agar is followed by incubation at 28°C for the isolation of mesophilic organisms (Babu et al., 2004).

2.2.1 Isolation of Extremophiles

For the isolation of marine thermopiles, muddy soil samples are collected from different sites of hot springs in sterile polybags and are brought immediately to the laboratory (Pandey et al., 2013). Further procedures are similar to that used for soil samples except with incubation at 50°C for 12 h in a nutrient broth. This can be continued with streaking on a nutrient agar and repeated testing of the isolates for their ability to withstand high temperatures. Hot springs, undersea hydrothermal vents, and volcanic lava can act as starting material for the isolation of hyperthermophiles such as *Pyrolobus fumarii* and *Methanopyrus kandleri* (Vieille and Zeikus, 2001). In contrast, true psychrophiles (at an optimum temperature 15°C or less) can be isolated from the Arctic or Antarctic environment and sea ice (Borriss et al., 2003).

The isolation of halophiles requires salt enrichment (10%–20% NaCl), which is followed by serial dilution and plating on a complete medium containing organic nutrients along with different salts, such as sulfates, phosphates, and chlorides, and trace metals

TABLE 2.1   Isolation of Extremophiles

Extremophile	Growth Conditions	Reference
Halophiles	Require at least 1 M salt for growth	Horikoshi and Bull (2011)
Thermophiles	Grow at temperatures between 50°C and 80°C	Vieille and Zeikus (2001)
Hyperthermophiles	Optimum growth at temperatures above 80°C	Vieille and Zeikus (2001)
Psychrophiles	Grow at temperatures between 10°C and 20°C	Junge et al. (2011)
Psychrotolerant	Grow at temperatures above 25°C, but also grow below 15°C	Junge et al. (2011)
Barophiles	Optimal growth pressure is more than 40 MPa	Li et al. (1998)

(Kumar and Karan, 2012). Samples are incubated at 30°C for 96 h. Repeated streaking results in isolation of pure cultures. They can be maintained by subculturing every 15 days and stored at 4°C. Halophilic organisms moderately exhibit a 5% salt requirement (Kushner, 1985). Hypersaline environments such as the Dead Sea, the Great Salt Lake, and solar salt evaporation ponds are found to be useful habitats for isolation of extreme halophilic microorganisms (salt requirement >25%) (Ventosa et al., 1998).

Sterilized mud samplers are used at a depth of more than 10,000 m for the isolation of barophilic microorganisms (Kato et al., 1998). The isolation procedure involves incubation under pressure of 100 MPa in plastic bags using marine broth, for example, strain MT41 (an extreme halophile that requires 700 atm pressure for optimum growth). However, in vitro culture of these microorganisms is difficult. Table 2.1 summarizes the growth conditions required for the isolation of extremophilic microorganisms.

2.2.2 Media

Microbial media are mainly of two types: defined or simple (containing inorganic nutrients) and organic or complex (containing organic nutrients). The type of medium selected for enrichment and growth depends on the type of microorganisms to be isolated. Different researchers have used different types of media for isolation purpose. A review of existing studies indicates modifications in conventional media. Although different media can be used for fungal isolation, typically used media are sabouraud agar, potato dextrose, glucose agar, malt agar, and so on. (Sengupta and Pramanik, 2015). Moreover, physical and chemical growth parameters may be varied in order to achieve maximum diversity. Eleven different types (IM1 to IM11) of selective media have been described by researchers, including starch casein agar, asparagine agar, oatmeal agar, yeast extract malt extract agar, and so on, for the isolation of marine microorganisms. Nutrient agar is also used for bacterial isolation followed by incubation at 28°C ± 2°C. Rose Bengal agar medium for fungal isolation and Kenknights agar medium for actinomycetes isolation were used by Lakshmipriya and Shivkumar (2012). Incubation conditions were 28°C for 2 days, 28°C for 3 days, and 30°C for 5–7 days. Usually, the plates are inoculated in triplicates for each dilution. The colonies obtained at the end of incubation period are subcultured based on their morphology. The isolates are further subjected to biochemical characterization and referred to *Bergys Manual of Determinative Bacteriology* for preliminary identification. Sequencing and phylogenetic analysis of 16S rDNA and 18S rDNA for bacteria and fungi, respectively, help to assign the isolates to specific genus and species. The isolates are routinely preserved on slants at 4°C, as

15% glycerol stocks at −80°C or by lyophilization. The choice of appropriate method depends on the isolate, its viability, and genetic stability.

2.3 SCREENING FOR INDUSTRIALLY IMPORTANT BIOACTIVE COMPOUNDS

2.3.1 Enzymes

Low cost, rapid, and sensitive methods for high-throughput screening of microbial enzymes are highly desirable. It is beneficial to add inducers in the medium at the isolation stage rather than including an additional screening test. This facilitates preferential selection of microorganisms that produce respective enzymes (Bonugli-Santos et al., 2015). Usually, detection methods consider color reaction or change in color of the media, for example, glutaminase and asperginase. Hydrolytic enzymes can be screened by plate assays using substrates, such as starch, gelatin, tributyrin, and so on, at 1% concentration for amylase, protease, and lipase production, respectively (Table 2.2). Here, one has to maintain the desired salinity in the medium and/or temperature of incubation depending on the sample used.

Starch agar plates are used for screening of amylase production. The plates at the end of the incubation period are treated with Gram's iodine and left for 5 min.

TABLE 2.2 Enzymes from Marine Microorganisms

Sr. No.	Enzyme	Screening Method	Source	Reference
1	Amylase	Starch iodine plate assay	Actinomycetes, Bacteria, Fungi	Selvam et al. (2011), Chimata et al. (2010), Shanmugasundaram et al. (2015)
2	Cellulase	Cellulolytic enzyme assay	Fungi	Elander (1987)
		Dilution pour plate technique in carboxymethyl cellulose agar medium	Bacteria	Premalatha et al. (2015)
3	Lipase	Rhodomine B agar plate assay	Actinomycetes	Selvam et al. (2011)
		Tributyrin nutrient agar plate assay	Bacteria	Vijayan et al. (2012)
		Simple qualitative plate assay		
		Tributyrin agar diffusion method	Fungi	Veerapagu et al. (2013)
4	Protease	Gelatin nutrient agar plates	Bacteria	Fulzele et al. (2011)
		Skim milk agar plates		
		Skim milk	Actinomycetes	Viswanathan et al. (2015)
5	ʟ-Asparaginase	ʟ-Asparaginase-glucose agar media	Bacteria	Jayadev et al. (2015)

The appearance of a clear zone against a blue background indicates amylase production (Imada and Simidu, 1988). Similarly, growth on a cellulose agar plate and the appearance of a clear zone with an iodine–potassium iodide solution against a dark background indicates cellulase production (Andro et al., 1984). Screening for lipase involves incubation in a Rhodamine B medium (Savitha et al., 2007). Colonies show orange fluorescence under UV light on the medium containing Rhodamine B and olive oil (Colen et al., 2006).

2.3.2 Enzyme Inhibitors

There are few papers available on enzyme inhibitors from marine bacteria or actinomycetes. Some inhibitors have therapeutic value, such as antihypertensive activity (Devi et al., 2010). Table 2.3 describes four such enzyme inhibitors along with their screening procedures.

TABLE 2.3   Enzyme Inhibitors

Inhibitor	Screening Method	Source	Reference
ACE-Angiotensin converting enzyme inhibitor	Preparation of agar containing ACE protein, placing filter paper discs impregnated with extract under test and overlaying with agar-containing substrate, and incubation at 37°C for 24 h followed by flooding of plates with 0.1 normal NaOH for 10 min; colorless zones on an amber background turning to pink after 30 min indicate a positive result	Bacteria	Wijesekara and Kim (2010)
ADA-Adenosine deaminase inhibitor	Use of agar plates containing adenosine deaminase and phenol red, placing filter paper discs impregnated with extract under test, incubation at room temp; yellow zones around the discs on purplish red background indicate a positive result	Bacteria	Chellaram and Prem Anand (2014)
α-Glucosidase inhibitor	Modified ρ-nitrophenyl-α-D-glucoside (PNPG) method and animal model experiments using male albino wistar rats	Actinomycetes	Ganesan et al. (2011)
Beta-glucosidase inhibitor	Inoculation of samples on enzyme agar plates and incubation at room temperature for 15 min and addition of 0.2% of esculin solution; appearance of pale yellowish zones against a blackish brown background at the end of incubation period of 30 min at room temperature indicates a positive result	Bacilli and actinomycetes	Pandey et al. (2013)

2.3.3 Exopolysaccharides

Marine microorganisms, particularly marine extremophiles (Biswas and Paul, 2014), produce high molecular weight heteropolysaccharides known as exopolysaccharides (EPS), which exist in the form of capsules, sheaths, or slime. They have a wide range of applications as thickeners, emulsifiers, and suspending agents in pharmaceutical, food, and other industries. In recent years, they have gained importance as immunostimulatory, antitumor, antiviral, and anti-inflammatory agents (Shankar et al., 2014). These polyanionic molecules are produced as a result of adaptations of microorganisms to harsh marine environmental conditions. EPSs play a significant role in the protection of cells from drying, toxic metals, phagocytosis, phage attack, and in formation of biofilms (Freitas et al., 2009). Screening of halophilic bacteria for EPS production was carried out by Biswas and Paul (2014) after isolation on malt extract-yeast extract medium (MY) medium. Saravanan and Jayachandran (2008) have used Zobell Marine Agar for different marine samples. In either case, mucoidal bacterial colonies or colonies with sticky consistency indicated EPS production. Other media used for EPS screening are BSS medium with 1% sucrose (Majumdar et al., 1990), marine agar medium supplemented with congo red (Fusconi and Godinho, 2002), and so on. Thus, an appearance of a mucoid colony is the first indication for EPS production, which can be further confirmed by repeated subculturing on media containing high sugar concentration.

2.3.4 Biosurfactants

Marine microorganisms are known to produce different types of biosurfactants, which are cationic, anionic, or zwitterionic compounds. Chemically, they are glycolipids, lipopolysaccharides, oligosaccharides, or lipoproteins with emulsification properties. These emulsifiers have various applications in the food, pharmaceutical, and cosmetic industries (Jadhav et al., 2013). They also play important roles in the bioremediation of oil spills and toxic pollutants. Recently, Antoniou et al. (2015) surveyed biosurfactants from marine microorganisms. There are nine different screening methods for the selection of biosurfactant-producing microbes, which include the following: hemolytic assay (Banat, 1993; Yonebayashi et al., 2000), bacterial adhesion to hydrocarbons (BATH) assay (Volchenko et al., 2007), drop collapse assay, oil spreading assay (Bodour and Miller-Maier, 1998; Youssef et al., 2004), emulsification assay (Afshar et al., 2008; Satpute et al., 2008), surface tension measurement, titled glass slide test, blue agar plate, and hydrocarbon overlay agar assay.

2.3.5 Polyunsaturated Fatty Acids

Polyunsaturated fatty acids (PUFA) are fatty acids that have more than one double bond in their backbone structure. Based on the position of their double bonds, they are grouped as ω-3, ω-6, and ω-9 fatty acids. ω-3 fatty acids (alpha linolenic acid [ALA], eicosapentaenoic acid [EPA], and docosahexaenoic acid [DHA]) and ω-6 fatty acids (linoleic acid [LA], gamma linolenic acid [GLA], and arachidonic acid [AA]) are of importance in human health (Jadhav et al., 2013; Pote et al., 2014), as various cardiovascular, cerebrovascular, autoimmune diseases, and cancers have been found to be associated with PUFA deficiency (Su, 2008). These long-chain fatty acids are produced by microorganisms to maintain membrane fluidity to survive in extreme habitats, mainly those with low temperatures and high pressures. Microbes are preferred over other PUFA sources like fish and algae owing to their (1) ability to produce single PUFA in high quantity, (2) renewable nature, (3) high oxidative stability, and (4) cost-effectiveness. Preliminary screening for PUFA production makes use of H_2O_2-plate assay. It involves placing of filter paper discs of diameter 5 mm impregnated with different concentrations of H_2O_2. Plates are then incubated at $28°C \pm 2°C$ for 24 h. The absence of an inhibition zone at the end of the incubation period indicates a positive result. Confirmation of PUFA production is based on Gas chromatography-Mass spectrometry (GC-MS) analysis of fatty acid methyl esters (FAMES) (Tilay and Annapure, 2012). For this, esterification of extracted lipid is carried out by different methods such as chloroform and methanol method (Chl:Met), dichloromethane and methanol method (Dic:Met), propan-2-ol and cyclohexane method (Pro:Hex), ethanol and KOH method (Eth:KOH), and ScCO$_2$: supercritical-CO$_2$ extraction method (Li et al., 2014).

2.3.6 Pigments

Large numbers of synthetic dyes are used as colorants in the food, textile, and pharmaceutical industries (Venil et al., 2013). However, there are certain drawbacks, such as hazardous effects on humans, animals, and the environment. In contrast, microbial pigments are found to have better biodegradability, stability, and environment acceptability. Hence, considerable economic potential lies in harvesting biocolorants for various industrial applications. The major classes of pigments include prodiginines, phenazine, quinones, tambjamines, and melanins, which are shown to have antibacterial, antifungal, anticancer, and immunosuppressive activities. The subclasses of these pigments, their microorganisms, structures, characteristics, and their applications were compiled by Soliev et al. (2011).

2.3.7 Marine Probiotics

Lactic acid bacteria isolated from dairy or nondairy foods mainly constitute a probiotic group of microorganisms. These microbes exert several health benefits to humans or the host when consumed in adequate quantities. However, the isolates have to satisfy certain criteria such as bile salt tolerance, acid tolerance, antimicrobial activity, adhesion to the human intestinal cell line, antibiotic susceptibility, and so on, to be considered as a probiotic (Dixit et al., 2013). To our knowledge, very few reports are available on probiotic microorganisms from the marine environment. Recently, Bernal et al. (2015) published a paper on marine actinomycetes wherein the isolates were examined for their resistance to low pH and bile salt. Their results indicated that the *Streptomyces* species LCJ94 could be used as a probiotic in aquaculture. The isolate also produced antibiotic and other bioactive compounds, which is an added advantage as it would help in reducing the risk of infection in aquaculture.

Screening of marine bacteria for probiotic characteristics, such as antibacterial activity against human pathogens and multiple antibiotic resistance, was conducted by Wadekar and Dharmadhikar (2015). The authors used the agar diffusion method on a Mueller–Hinton agar to examine the antimicrobial activity of a cell-free broth of the isolates that were previously grown in a DeMan, Rogosa, and Sharp (MRS) broth. The zones of inhibition at the end of incubation at 37°C indicated their broad spectrum antibacterial activity.

2.4 SCREENING FOR BIOMEDICALLY IMPORTANT COMPOUNDS

The biomedical activities of microbial metabolites described next are listed in Table 2.4.

2.4.1 Antibacterial Activity

The increase in multiple drug resistance in bacteria is one of the major reasons for the search for novel molecules with antibacterial activity. The accomplishments are reviewed by Jensen and Fenical (1994) and Bernan et al. (1997). Marine actinomycetes isolated from sea sediments, plants, or animals are the major source of microorganisms producing such compounds (Zheng et al., 2000). The standard procedure of Kirby–Bauer disc diffusion assay for screening of antibacterial activity makes use of the paper disc assay against test microorganisms. Here, paper discs saturated with the extracts under examination are placed on an agar surface seeded with test microorganisms. The diameter of the inhibition zone at the end of the incubation period indicates antimicrobial activity

TABLE 2.4   Screening for Biomedical Activity

Activity	Microorganisms	Compound	Screening	Reference
Antioxidant	Bacteria	Exopolysaccharides Crude extract	DPPH assay, Cupric ion reducing antioxidant capacity (CUPRAC) assay	Selim et al. (2015)
	Actinomycetes	Ethyl acetate (EA) extract	ABTS assay	
Antimicrobial	Actinomycetes	Ethyl acetate (EA) extract	Cup-well agar diffusion method, block and well diffusion method	Priya et al. (2012)
	Bacteria	Crude extract	Agar diffusion test	Al-Zereini (2014)
Anticancer	Bacteria	Crude extract	MTT assay	Utami et al. (2014)
Antitumor	Actinomycetes	Crude antibiotic Fermentation broth	Cytotoxic assay MTT assay	Pham et al. (2014)
Antifungal	Actinomycetes	Ethyl acetate extract	Cross streak method	Priya et al. (2012)
Anti-inflammatory	Fungi	Pyrenocine A	Anti-inflammatory activity assays using mouse leukemic monocyte-macrophage cell line	Toledo et al. (2014)

(Kumari et al., 2013). Other methods include the cross-streak method (Saxena et al., 2013) or agar well diffusion method. A more elaborate technique of autobiography overlay assay was used by Jeganathan et al. (2013) to evaluate the antimicrobial activity of marine bacteria. The compounds from marine actinomycetes and bacteria are generally screened for various human pathogens.

2.4.2 Antifungal Activity

During the past decade, there has been a significant increase in life-threatening infections caused by fungi (Rapp, 2004). They are mainly caused by the *Candida*, *Aspergillus*, and *Fusarium* spp. in immune-compromised patients or patients with chronic diseases, such as AIDS. Therefore, effective antifungal compounds with selective toxicity are always the need of the hour. In this context, researchers have documented bioactive metabolites from bacteria and actinomycetes with antagonistic activity against pathogenic fungi (Park et al., 2008; Kong et al., 2010). The extracts are generally screened using the disc diffusion assay method. The sterile paper discs impregnated with extracts are used to examine antifungal activity. The plates are incubated for 48 h and appearance of a zone of inhibition around the disc is considered as a positive result (Devi et al., 2010).

2.4.3 Anti-HIV Activity

Marine microorganisms are also found to be an excellent source of bioactive compounds with anti-HIV activity. AIDS caused by the human immunodeficiency virus (HIV) is a life-threatening disease and most deaths occur due to secondary infections resulting from loss of immunity. Different anti-HIV drugs target different stages in the life cycle of HIV, for example, reverse transcriptase (RT) inhibitors, protease (PR) inhibitors, integrase (IN) inhibitors, and so on. However, the increase in drug resistance and adverse side effects of these compounds has led scientists to look into marine resources for antiretroviral therapy (Gochfeld et al., 2003). In 2013, Zhou et al. published a review on 132 marine-derived anti-HIV constituents along with their sources, targets, and structure–function relationship. Out of 150 natural metabolites reported by Tziveleka et al. (2003), equisetin, phomasetin, and integric acid produced by marine fungi have shown anti-HIV activity. The procedure for screening for anti-HIV activity differs with respect to the target molecule.

2.4.4 Anticancer Activity

Cytarabine, the synthetic anticancer drug based on C-nucleosides isolated from marine sponge, is used for the treatment of leukaemia and lymphoma (Cragg and Newman, 1999; Schwartsmann et al., 2001). Since this discovery, different anticancer agents obtained from marine sources are under preclinical and clinical trials, for example, bryostatins, discodermolide, eleutherobin, and sarcobictyin (Singh et al., 2008). The review published by Schwartsmann et al. (2001) summarizes the evaluation of these novel compounds undergoing phase I and II trials. Hussain et al. (2012) also emphasized on pharmacologically active marine natural products with cytotoxic, anticancer, antitopoi-somerase-I, and in vivo activity. Generally, the anticancer activity of a desired product/extract is evaluated by 3-(4,5-Dimethylthiazol-2-yl)-2,5-Diphenyltetrazolium Bromide (MTT) assay for which serial dilutions of DMSO dissolved samples are used to examine anti-proliferative effect on cancerous cell lines, such as the human epithelial carcinoma cell line (HeLa). Incubation for 24 h at 37°C is followed by the addition of MTT, incubation at 37°C for 4 h, and absorbance measurement by enzyme-linked immunosorbent assay (ELISA) reader at 595 nm. The relationship between the concentration of the extract and % inhibition helps to determine IC_{50} value (Rahman and Hussain, 2015).

2.4.5 Antioxidant Activity

The studies on marine seaweed, sponges, algae, and microorganisms indicate that they are rich sources of antioxidants. These bioactive compounds can be used to develop antiaging and anticarcinogenic products for human health benefits (Balakrishnan et al., 2014). The antioxidant capacity of crude extracts from marine microbes can be measured using different types of assays such as 1,1-diphenyl-2-picryl-hydrazyl (DPPH) scavenging activity, hydrogen peroxide scavenging (H_2O_2) assay, nitric oxide scavenging activity, trolox equivalent antioxidant capacity (TEAC) method/ABTS radical cation decolorization assay, total radical-trapping antioxidant parameter (TRAP) method, ferric reducing-antioxidant power (FRAP) assay, superoxide radical scavenging activity (SOD), hydroxyl radical scavenging activity, hydroxyl radical averting capacity (HORAC) method, oxygen radical absorbance capacity (ORAC) method, reducing power method (RP), ferric thiocyanate (FTC) method, thiobarbituric acid (TBA) method, DMPD (N,N-dimethyl-p-phenylene diamine dihydrochloride) method, xanthine oxidase method, cupric ion reducing antioxidant capacity (CUPRAC) method, and metal chelating activity (Alam et al., 2013).

2.5 CONCLUSION

Marine microorganisms are a rich source of biological and chemical diversity that produce a variety of bioactive compounds with unique structural features. The studies on bioactive compounds have not only resulted in the development of useful drugs or drug candidates but also chemical compounds with potential applications in cosmetics, nutritional supplements, molecular probe, agrochemicals, and so on. It is worth mentioning that more than 60% of the anticancer drugs commercially available are of natural origin, and many new compounds are being studied every year (Rehm, 2009, 2010). With great improvements in deep sea sample collection devices and technologies, it will be possible to collect a wide range of plants and animals that would help in isolating microorganisms. Recent advancements in the techniques in molecular biology will definitely help in better understanding of secondary metabolism in fungi and bacteria. Detailed knowledge of the corresponding gene clusters and the factors affecting their expression will prove useful in more intensive investigations of the chemical and pharmacological properties of these bioactive molecules. Although marine ecosystem is a rich source of diverse microorganisms, very few are culturable in vitro, and hence developing new isolation methods for marine-derived microbes is of paramount importance.

REFERENCES

Afshar, S., Lotfabad, T.B., and Roostaazad, R. Comparative approach for detection of biosurfactant-producing bacteria isolated from Ahvaz petroleum excavation areas in south of Iran. *Ann. Microbiol.*, 58 (2008): 555–560.

Alam, M.N., Bristi, N.J., Rafiquzzaman, M. et al. Review on in vivo and in vitro methods evaluation of antioxidant activity. *Saudi Pharm. J.*, 21 (2013): 143–152.

Al-Zereini, W.A. Bioactive crude extracts from four bacterial isolates of marine sediments from Red Sea, Gulf of Aqaba, Jordan. *J. Biol. Sci.*, 7(2) (2014): 133–137.

Andro, T., Chambost, J.P., Kotoujansky, A. et al. Mutants of *Erwinia chrysanthemi* defective in secretion of pectinase and cellulose. *J. Bacteriol.*, 160 (1984): 1199–1203.

Antoniou, E., Fodelianakis, S., Korkakaki, E., and Kalogerakis, N. Biosurfactant production from marine hydrocarbon-degrading consortia and pure bacterial strains using crude oil as carbon source. *Front. Microbiol.*, (2015): 00274.

Babu, T.G., Nithyanand, P., Kannapiran, E. et al. Molecular identification of bacteria associated with the coral reef ecosystem of Gulf of Mannar Marine Biosphere Reserve using 16S rRNA sequences. *Front. Mar. Biosci. Res.*, (2004): 47–53.

Balakrishnan, D., Kandasamy, D., Nithyanand, P. et al. A review on antioxidant activity of marine organisms. *Int. J. ChemTech Res.*, 6(7) (2014): 3431–3436.

Banat, I.M. The isolation of a thermophilic biosurfactant producing *Bacillus* sp. *Biotechnol. Lett.*, 15 (1993): 591–594.

Bernal, M.G., Campa-Córdova, A.I., Saucedo, P.E. et al. Isolation and in vitro selection of actinomycetes strains as potential probiotics for aquaculture. *Veter. World* (2015).

Bernan, V.S., Greenstein, M., Maiese, W.M. et al. Marine microorganisms as a source of new natural products. *Adv. Appl. Microbiol.*, 43 (1997): 57–90.

Biswas, J. and Paul, A.K. Production of extracellular polymeric substances by halophilic bacteria of solar salterns. *Chin. J. Biol.*, (2014), Article ID 205731: 12 pp.

Bodour, A.A. and Miller-Maier, R. Application of a modified dropcollapse technique for surfactant quantification and screening of biosurfactant-producing microorganisms. *J. Microbiol. Methods*, 32 (1998): 273–280.

Bonugli-Santos, R.C. and dos Santos Vasconcelos, M.R., Passarini, M.R. et al. Marine-derived fungi: Diversity of enzymes and biotechnological applications. *Front. Microbiol.*, 6 (2015): 269.

Borriss, M., Helmke, E., Hanschke, R. et al. Isolation and characterization of marine psychrophilic phage-host systems from Arctic sea ice. *Extremophiles*, 7(5) (2003): 377–384.

Chellaram, C. and Prem Anand, T. Screening for enzyme inhibitors in marine bacteria. *Int. J. PharmTech Res.*, 6(1) (2014): 351–355.

Chimata, M.K., Sasidhar, P., Challa, S. et al. Production of extracellular amylase from agricultural residues by a newly isolated *Aspergillus* species in solid state fermentation. *Afr. J. Biotechnol.*, 9(32) (2010): 5162–5169.

Colen, G., Junqueira, R.G., and Moraes-Santos, T. et al. Isolation and screening of alkaline lipase producing fungi from Brazilian Savanna soil. *Microbiol. Biotechnol.*, 22 (2006): 881–885.

Cragg, G.M. and Newman, D.J. Discovery and development of antineoplastic agents from natural sources. *Cancer Invest.*, 17(1999): 153–163.

da Silva, M.A., Cavalett, A., Spinner, A., Rosa, D.C., Jasper, R.B., Quecine, M.C., Bonatelli, M.L., Pizzirani-Kleiner, A., Corção, G., and Lima, A.O. Phylogenetic identification of marine bacteria isolated from deep-sea sediments of the eastern South Atlantic Ocean. *SpringerPlus*, 2 (2013): 127.

Devi, P., Wahidullah, S., and Rodrigues, C. The sponge-associated bacterium *Bacillus licheniformis* SAB1: A source of antimicrobial compounds. *Mar. Drugs*, 8(4) (2010): 1203–1212.

Dixit, G., Samarth, D., Tale, V., and Bhadekar, R. Comparative studies on potential probiotic characteristics of *Lactobacillus acidophilus* strains. *Eurasia J. Biosci.*, 7 (2013): 1–9.

Dhevendaran, K. and Anithakumarai, K. L-asparaginase activity in growing conditions of *Streptomyces* sp., associated with *Therapon jarbuo* and Villiorita Cuprinoids of Veli Lake, South India. *Fish Technol.*, 39 (2002): 155–159.

Elander, R.P., Microbial screening, selection and strain improvement, In: *Basic Biotechnology*, Lock, J.B. and Krintianasen, B., Eds. Academic Press, London, UK, 1987, pp. 271–251.

Foucher, JP. The future of integrated deep-sea research in Europe. *Oceanography*, 22(1) (2009): 178.

Fulzele, R., DeSa, E., Yadav, A. et al. Characterization of novel extracellular protease produced by marine bacterial isolate from the Indian Ocean. *Braz. J. Microbiol.*, 42(4) (2011): 1364–1373.

Freitas, F., Alves, V.D., Pais, J. et al. Characterization of an extracellular polysaccharide produced by a *Pseudomonas* strain grown on glycerol. *Bioresour. Technol.*, 100 (2009): 859–865.

Fusconi, R. and Godinho, M. Screening for exopolysaccharide-producing bacteria from sub-tropical polluted groundwater. *Braz. J. Biol.*, 62(2) (2002): 363–369.

Ganesan, S., Raja, S., Sampathkumar, P. et al. Isolation and screening of α-glucosidase enzyme inhibitor producing marine actinobacteria. *Afr. J. Microbiol. Res.*, 5(21) (2011): 3437–3245.

Gochfeld, D.J, Sayed, K.A.E, Yousaf, M. et al. Marine natural products as lead anti-HIV agents. *Curr. Med. Chem.*, (2003): 401–424.

Hong, K., Gao, A.H., Xie, Q.Y., Gao, H., Zhuang, L., Lin, H.P., Yu, H.P., Li, J., Yao, X.S., Goodfellow, M., and Ruan, J.S. Actinomycetes for marine drug discovery isolated from mangrove soils and plants in China. *Mar. Drugs*, 7(1) (2009): 24–44.

Horikoshi, K. and Bull, A.T. Prologue: Definition, categories, distribution, origin and evolution, pioneering studies, and emerging fields of extremophiles. In: *Extremophiles Handbook*. Horikoshi, K., Antranikaian, G., Bull, A.T., Robb, F.T., and Stetter, K.O., Eds. Springer, Tokyo, Japan, 2011, pp. 4–15.

Hussain, M.S., Fareed, S., Saba Ansari, M. et al. Marine natural products: A lead for Anti-cancer. *Indian J. Geo-Mar. Sci.*, 41(1) (2012): 27–39.

Imada, C. and Simidu, U. Isolation and characterization of an α-amylase inhibitor producing actinomycetes from marine environment. *Nippon Suisan Gakkaishi*, 54(10) (1988): 1839–1845.

Jadhav, V.V., Arora, A., Bhadekar, R.K. et al. Studies on biosurfactant from *Oceanobacillus* sp. BRI 10 isolated from Antarctic sea water. *Desalination*, 318 (2013): 64–71.

Jayadev, J., Lekshmi, M., Sreelekshmi, V. et al. Marine bacteria: A potential bioresource for multiple applications. *Int. J. Sci. Eng. Res.*, 6(9) (2015).

Jeganathan, P., Rajasekaran, K.M., Asha Devi, N.K. Antimicrobial activity and characterization of marine bacteria. *Indian J. Pharm. Biol. Res.*, 1(4) (2013): 38–44.

Jensen, P.R. and Fenical, W. Strategies for the discovery of secondary metabolites from marine bacteria: Ecological perspectives. *Annu. Rev. Microbiol.*, 48 (1994): 559–584.

Junge, K., Christner, B., Staley, J.T. et al. Diversity of psychrophilic bacteria from sea ice—And glacial ice communities. In: *Extremophiles Handbook*, Springer, Japan, 2011, pp. 793–815.

Kato, C., Li, L., Nogi, Y. et al. Extremely barophilic bacteria isolated from the Mariana Trench, Challenger Deep, at a depth of 11,000 meter. *Appl. Environ. Microbiol.*, 64(4) (1998): 1510–1513.

Kong, Q., Shan, S., Liu, Q. et al. Biocontrol of *Aspergillus flavus* on peanut kernels by use of a strain of marine *Bacillus megaterium. Int. J. Food Microbiol.*, 139(1) (2010): 31–35.

Kumar, S. and Karan, R. Screening and isolation of halophilic bacteria producing industrially important enzymes. *Braz. J. Microbiol.*, 43(4) (2012): 1595–1603.

Kumari, C.K., Madhuri, R.J., and Reddy, C.D. Antimicrobial potential of marine bacterial isolates from different coastal regions of Andhra Pradesh and Tamil Nadu, India. *Int. J. Curr. Microbiol. Appl. Sci.*, 2(10) (2013): 230–237.

Kushner, D.J. The Halobacteriaceae. In: *The Bacteria A Treatise on Structure and Function*. Woese, C.R. and Wolfe, R.S., Eds. Vol. 8, Academic Press, New York, 1985, pp. 171–214.

Lakshmipriya, V.P. and Sivakumar, P.K. Isolation and characterization of total heterotrophic bacteria and exopolysaccharide produced from mangrove ecosystem. *Int. J. Pharm. Biol. Arch.*, 3(3) (2012): 679–684.

Li, L., Kato, C., Nogi, Y. et al. Distribution of the pressure-regulated operons in deep-sea bacteria. *FEMS Microbiol. Lett.*, 159(2) (1998): 159–166.

Li, Y., Naghdi, F.G., Garg, S. et al. A comparative study: The impact of different lipid extraction methods on current microalgal lipid research. *Microb. Cell Fact.*, (2014): 13–14.

Majumdar, I., D'soula, F., and Bhosle, N.B. Microbial exopolysaccharides: Effect on corrosion and partial chemical characterization. *J. Ind. Inst. Sci.*, 79 (1990): 539–550.

Newel, S.Y. Mangrove fungi: The succession in the mycroflora of red mangrove (*Rhizophora mangle* L.). In: *Recent Advances in Aquatic Mycology*. Jones, E.B.G., Ed. Paul Elek Scientific Books, London, UK, 1976.

Nybakken, J.W. and Bertne, M.D. Comparison of terrestrial and marine ecosystems. In: *Marine Biology: An Ecological Approach*, Pearson, USA, 2004, pp. 28–32.

Pandey, S., Sree, A., Dash, S.S., and Sethi, D.P. A novel method for screening beta-glucosidase inhibitors. *BMC Microbiol.*, 13 (2013): 55.

Park, C.N., Lee, J.M., Lee, D. et al. Antifungal activity of valinomycin, a peptide antibiotic produced by *Streptomyces* sp. strain M 10 antagonistic to *Botrytis cinerea. J. Microbiol. Biotechnol.*, 18(5) (2008): 880–884.

Pham, H.T., Nguyen, N.P., Phi, T.Q., Dang, P.T., and Le, H.G. The antibacterial and anticancer activity of marine Actinomycete strain HP411 isolated in the Northern Coast of Vietnam. *Int. J. Biotechnol. Bioeng.*, 1(12) (2014).

Pimpliskar, M.R. and Jadhav, R.N. Isolation of marine bacteria along Vasai Coast (M.S.) India. *Int. J. Curr. Res. Acad. Rev.*, 2(4) (2014): 89–93.

Pote, S. and Bhadekar, R. Statistical approach for production of PUFA from *Kocuria* sp. BRI 35 isolated from marine water sample. *Biomed. Res. Int.*, (2014): 570925.

Premalatha, N., Gopal, N.O., Jose, P.A. et al. Optimization of cellulase production by *Enhydrobacter* sp. ACCA2 and its application in biomass saccharification. *Front. Microbiol.*, 6(1046) (2015).

Priya, J., Sagadeva, E., Dhanalakshmi, P. et al. Detection of antioxidant and antimicrobial activities in marine actinomycetes isolated from Puducherry Coastal Region Arulappan. *J. Modern Biotechnol.*, 1(2) (2012): 63–69.

Rahman, A. and Hussain, A. Anticancer activity and apoptosis inducing effect of methanolic extract of *Cordia dichotoma* against human cancer cell line. *Bangladesh J. Pharmacol.*, 10 (2015): 27–34.

Rapp, R.P. Changing strategies for the management of invasive fungal infections. *Pharmacotherapy*, 24 (2004): 4S–28S.

Rehm, B. *Microbial Production of Biopolymers and Polymer Precursors: Applications and Perspectives.* Caister Academic, Norfolk, UK, 2009.

Rehm, B.H. Bacterial polymers: Biosynthesis, modifications and applications. *Nat. Rev. Microbiol.*, 8 (2010): 578–592.

Saravanan, P. and Jayachandran, S. Preliminary characterization of exopolysaccharides produced by a marine biofilm-forming bacterium *Pseudoalteromonas ruthenica* (SBT 033). *Lett. Appl. Microbiol.*, 46 (2008): 1–6.

Satpute, S.K., Bhawsar, B.D., Dhakephalkar, P.K. et al. Assessment of different screening methods for selecting biosurfactant producing marine bacteria. *Indian J. Mar. Sci.*, 37 (2008): 243–250.

Savitha, J., Srividya, S., Jagat, R. et al. Identification of potential fungal strains for the production of inducible, extracellular and alkalophilic lipase. *Afr. J. Biotechnol.*, 6 (2007): 564–568.

Saxena, A., Upadhyay, R., Kumar, D. et al. Isolation, antifungal activity and characterization of soil actinomycetes. *J. Sci. Ind. Res.*, 72, August (2013): 491–497.

Schwartsmann, G., Brondani da Rocha, A., and Berlinck, R.G.S. Marine organisms as a source of new anticancer agents. *Lancet Oncol.*, 2 (2001): 221–225.

Selim, M.S., Mohamed, S.S., Shimaa, R.H. et al. Screening of bacterial antioxidant exopolysaccharides isolated from Egyptian habitats. *J. Chem. Pharm. Res.*, 7(4) (2015): 980–986.

Selvam, K., Vishnupriya, B., and Subhash Chandra Bose, V. Screening and quantification of marine actinomycetes producing industrial enzymes amylase, cellulase and lipase from south coast of India. *Int. J. Pharm. Biol. Arch.*, 2(5) (2011): 1481–1487.

Shankar, T., Vijayabaskar, P., Sivasankara, N. et al. Screening of exopolysaccharide producing bacterium *Frateuria aurentia* from elephant dung. *App. Sci. Rep.*, 5(3) (2014): 105–109.

Shanmugasundaram, S., Eswar, A., Mayavu, P., and Surya, M. Screening and identification of amylase producing bacteria from Marakkanam Saltpan Environment, Tamil Nadu, India. *Asian J. Biomed. Pharm. Sci.*, 5(48) (2015): 35–37.

Singh, R., Sharma, M., Joshi, P. et al. Clinical status of anti-cancer agents derived from marine sources. *Anticancer Agents Med. Chem.*, 8(6) (2008): 603–617.

Soliev, A.B., Hosokawa, K., and Enomoto, K. Bioactive pigments from marine bacteria: Applications and physiological roles. *Evid Based Complement. Altern. Med.*, (2011): 17.

Sengupta, S. and Pramanik, A. Antimicrobial activities of actinomycetes isolated from unexplored regions of Sundarbans mangrove ecosystem. *BMC Microbiol.*, 15 (2015): 170.

Su, K.P. Mind-body interface: The role of n-3 fatty acids in psychoneuroimmunology, somatic presentation, and medical illness comorbidity of depression. *Asia Pacific J. Clin. Nutr.*, 17(1) (2008): 151–157.

Tilay, A. and Annapure, U. Novel simplified and rapid method for screening and isolation of polyunsaturated fatty acids producing marine bacteria. *Biotechnol. Res. Int.* (2012): 8.

Toledo, T.R., Dejani, N.N., Silva Monnazzi, L.G. et al. Potent anti-inflammatory activity of pyrenocine A isolated from the marine-derived fungus *Penicillium paxilli* Ma(G)K. *Mediators Inflamm.*, 2014 (2014): 11.

Tziveleka, L.A., Vagias, C., Roussis, V. Natural products with anti-HIV from marine organisms. *Curr. Top. Med. Chem.*, 1 (2003): 1512–1535.

Utami, A.W.A. Wahyudi, A.T., and Batubara, I. Toxicity, anticancer and antioxidant activity of extracts from marine bacteria associated with sponge. *Int. J. Pharm. Bio. Sci.*, 5(4) (2014): 917–923.

Vijayan, E., Sagadevan, E., Arumugam, P. et al. Screening of marine bacteria for multiple biotechnological applications. *J. Acad. Indus. Res.*, 1(6) (2012): 5.

Venil, C.K., Zakaria, Z.A., and Ahmad, W.A. Bacterial pigments and their applications process. *Biochemistry*, 48(7) (2013): 1065–1079.

Veerapagu, M., Sankara Narayanan, A., Ponmurugan, K. et al. Screening selection identification production and optimization of bacterial lipase from oil spilled soil. *M Asian J. Pharm. Clin. Res.*, 6(3) (2013): 62–67.

Ventosa, A., Nieto, J.J., and Oren, A. Biology of moderately halophilic aerobic bacteria. *Microbiol. Mol. Biol.*, 62(2) (1998): 504–544.

Vieille, C. and Zeikus, G.J. Hyperthermophilic enzymes: Sources, uses, and molecular mechanisms for thermostability. *Microbiol. Mol. Biol.*, 65(1) (2001): 1–43.

Viswanathan, L., Rebecca, J., Arumugam, P. et al. Isolation and screening of protease producing marine actinomycetes from Chennai coastal region. *Int. J. Adv. Res. Biol. Sci.*, 2(8) (2015): 153–157.

Volchenko, N.N., Karasev, S.G., Nimchenko, D.V. et al. Cell hydrophobicity as a criterion of selection of bacterial producers of biosurfactants. *Microbiology*, 76 (2007): 112–114.

Wadekar, B.P. and Dharmadhikar, S.M. Isolation and screening of marine bacteria as probiotics and evaluation of antibacterial activity against human pathogens. *Indian J. L. Sci.*, 4(2) (2015): 79–86.

Wijesekara, I. and Kim, S.K. Angiotensin-I-converting enzyme (ACE) inhibitors from marine resources: Prospects in the pharmaceutical industry. *Mar. Drugs*, 8(4) (2010): 1080–1093.

Yonebayashi, H., Yoshida, S., Ono, K. et al. Screening of microorganisms for microbial enhanced oil recovery process. *Sekiyu Gakkaishi*, 43 (2000): 59–69.

Youssef, N.H., Duncan, K.E., and Nagle, D.P. Comparison of methods to detect biosurfactant production by diverse microorganisms. *J. Microbiol. Methods*, 56 (2004): 339–347.

Zheng, Z., Zeng, W., Huang, Y. et al. Detection of antitumor and antimicrobial activities in marine organism associated actinomycetes isolated from the Taiwan Strait, China. *FEMS Microbiol. Lett.*, 188(1) (2000): 87–91.

Zhou, X., Liu, J., Yang, B. et al. Marine natural products with anti-HIV activities in the last decade. *Curr. Med. Chem.*, 20(7) (2013): 953–973.

CHAPTER **3**

Extraction, Composition, and Quantification of Carbohydrates

Ompal Singh, H. S. Rathore, and Leo M. L. Nollet

CONTENTS

3.1	Marine Microorganisms	36
	3.1.1 Exopolymeric Substances	37
3.2	Carbohydrates	37
	3.2.1 Characteristics of Sugars	38
3.3	The Composition of Marine Microorganism Carbohydrates	39
3.4	Analysis of Carbohydrates	39
3.5	Reactive Sites of Carbohydrates	40
3.6	Detection of Carbohydrates in the Field	40
	3.6.1 Testing through Transformation to Furfural	40
	3.6.2 Testing with Triphenyltetrazolium Chloride	41
	3.6.3 Testing with 5-Hydroxy-1-Tetralone	41
	3.6.4 Testing with Stannous Chloride, Sulfuric Acid, and Urea	42
3.7	Isolation and Purification Methods of Extracellular Polymeric Substances	42
	3.7.1 Common Methods of Extraction	42
	3.7.1.1 Dialysis or Diafiltration	42
	3.7.1.2 Precipitation	43
	3.7.1.3 Enzyme Digestion	43
	3.7.1.4 Size Exclusion Chromatography	43
	3.7.1.5 Anion Exchange Chromatography	43
	3.7.1.6 Single-Dimension Gel Electrophoresis	43
	3.7.1.7 Two-Dimensional SDS-PAGE	44
	3.7.2 Ultrasensitive, Sophisticated, and Costly Methods	44
	3.7.2.1 Supercritical Fluid Extraction	44
	3.7.2.2 Pressurized Liquid Extraction	45
	3.7.2.3 Pressurized Hot Water Extraction	45
	3.7.2.4 New Approaches	45
3.8	Quantification of Marine Microorganism Carbohydrates	46
	3.8.1 Colorimetric/Spectrophotometric Method	46
	3.8.2 Chromatography	47
	3.8.2.1 Paper Chromatography	47
	3.8.2.2 Thin-Layer Chromatography	47

3.8.2.3 Low-Pressure Liquid Chromatography   48
3.8.2.4 High-Pressure Liquid Column Chromatography   48
3.8.2.5 Gas Chromatography-Flame Ionization Detection   49
3.8.3 Mass Spectrometry   50
3.8.4 Enzymatic Methods   50
3.8.5 Other Techniques   51
References   51

3.1 MARINE MICROORGANISMS

Marine microbes are tiny organisms that live in marine environment and can only be seen under a microscope. They include cellular life forms—bacteria, fungi, algae, and plankton—along with the viruses that freeload on the cellular life forms. Curtis Suttle, University of British Columbia, defines viruses as "Viruses are the most abundant life form in the oceans... and if stretched end to end, would span further than the nearest 60 galaxies." There are more than a billion microorganisms living in each liter of seawater, and it is now known that microbes dominate the abundance, diversity, and metabolic activity of the ocean. They comprise 98% of the biomass of the world's ocean, supply more than half the world's oxygen, are the major processors of the world's greenhouse gases, and have the potential to migrate the effects of climate change. Scientists are only just beginning to understand the important environmental roles that microbes play in marine systems—from feeding ecosystems to consuming waste and sequestering carbon. The Australian Institute of Marine Science is investigating several areas where microbial processes are central to the issue of immediate concern for the world's coral reefs. They are the cause of diseases that are suspected to be spreading due to global warming, yet paradoxically these compounds produce potential cancer cures and solutions for combating human disease. AIMS scientists have the ability to comprehensively study these compounds, by extracting them and analyzing their structure and their effects on mammalian cells. Research focus areas will include studies of the symbiotic and pathogenic relationships between marine microbes and other marine organisms.

Over the past decade (Bernan et al., 1997), marine microorganisms have become recognized as an important and untapped resource for novel bioactive compounds. The oceans cover greater than 70% of the earth's surface and, taking this fact into account by volume, represent more than 95% of the biosphere. Considering this fact, the oceans present themselves as an unexplored area of opportunity for the discovery of pharmacologically active compounds. However, it is important to pursue basic researches on the marine environment in order to permit the isolation of bioactive compounds (Gilmour et al., 2013; English et al., 2013). In this chapter, data have been presented to illustrate the possibility of detection (semiqualitative analysis), extraction (enrichment and cleanup), and quantification (estimation/determination) of carbohydrates in marine microorganisms. It is obvious that a greater investment in the development of marine biotechnology will produce novel compounds that may contribute significantly toward drug development over the next decade.

Macroalgae are regarded as a rich source of sulfated polysaccharides and the particular type of polysaccharide is different depending on taxonomic group (Ibanez et al., 2012). Different carbohydrates including agar, carrageenan, or alginates are extracted from macroalgae, and these carbohydrates are used widely in the food and pharmaceutical industries as functional ingredients such as stabilizers; that is, the alga *Chondrus crispus* is traditionally employed for the extraction of carrageenan (Irish moss), a highly sulfated polysaccharide. The macroalgal polysaccharides also have a potential for use as prebiotic as they are not digested in the human gut. They can be considered as a rich source of dietetic fiber. They also possess the associated bioactive properties: immunomodulating, anticancer, anti-inflammatory, antiviral, or antioxidant activities. Other minor sulfated polysaccharides (porphyrans produced by *Porphyra*) are also known. Generally, these polysaccharides may differ in their composition, and in turn, their properties are also differing; that is, the bioactivity depends on the degree of sulfation, molecular weight, type of dominating sugar, and glycoside branching. Chitin is another product that is very similar to carbohydrate. Chitin is one of the most extensive biopolymers in nature, which is abundant in different marine sources such as crustaceans, where it is part of their exoskeleton. Chitin can be converted in chitosan by the alkaline deacetylation of chitin and employed in a wide range of applications. It is associated with some interesting effects such as dietary fiber, lipid absorption reduction, and hypocholesterolemic or antidiabetic effects.

3.1.1 Exopolymeric Substances

Bacterial-exopolymeric substances (EPS) are molecules released in response to physiological stress (low nutrient concentrations; high concentration of toxins, metals, and pH; and extreme salinity) encountered in the natural environment. EPS may be defined as "they are high molecular weight compounds secreted by microorganisms into their environment." Exoenzymes attached to the cell surface can effectively hydrolyze large foreign molecules into more readily utilizable molecules, for example, amino acids and monosaccharides, to be utilized by the bacteria itself. These polymers have been characterized as a heterogeneous mixture of polysaccharides, proteins, with minor amount of lipids, nucleic acids, and other polymers such as flagella, phages, debris from lased cell, outer membrane vesicles, and pili. The densely packed and less diffusible capsules with a more organized polymeric structure are called "capsular" EPS or "attached" EPS. The polymers that more loosely adhere in the form of slime and can more easily be shed into the extracellular medium, as well as those that are already free in the medium, are called "nonattached" EPS. Various components of EPS are carbohydrates, acid polysaccharides (APS), uronic acid, proteins, and nucleic acids.

3.2 CARBOHYDRATES

As indicated by their name "carbohydrates" consist of, besides carbon, hydrogen and oxygen in the ratio of 2:1 as in water. Their general formula is $Cx(H_2O)y$, so they seem to be hydrates of carbon. Hence, x and y are whole numbers, and if x is between three and seven, such carbohydrates are called simple sugars or monosaccharides. The monosaccharides are further classified on the basis of carbon atoms in their molecules: C3 sugars are called trioses; C4 sugars, tetroses; C5 sugars, pentoses; C6 sugars, hexoses; and C7 sugars, heptoses. The suffix *-ose* is used to denote sugars.

The monosaccharides are polyhydroxy compounds and are either aldehydes or ketones. In aldehydic sugars, the C=O group is the terminal group, and in keto sugars, it takes a position next to the terminal carbon. The aldehydic sugars are collectively called aldoses and the ketonic sugars, ketoses. The common aldoses are ribose (C5 sugar) and glucose (C6 sugar) and the most common ketose is fructose (C6 sugar). In chemical oceanography, the term "neutral sugars" or neutral carbohydrates is synonymous to neutral monosaccharides (Panagiotopoulos and Sempere, 2005a) including aldohexoses (glucose, galactose, and mannose), aldopentoses (arabinose, xylose, and ribose), and deoxy sugars (fucose and rhamnose). Typical examples of some monosaccharides are aldohexoses, deoxy sugars, aldopentoses, amino sugars, ketohexoses, uronic acids, and so on. These comprise the majority of carbohydrate building blocks and intracellular metabolites. They are commonly released after acid hydrolysis of polysaccharides.

3.2.1 Characteristics of Sugars

All monosaccharides have asymmetric carbon atoms and therefore exhibit optical activity. Most of the naturally occurring sugars belong to the D-series and exhibit a property called mutarotation. Mutarotation is the property of a sugar showing optical rotations under different conditions. This has been attributed to the existence of sugar molecules in ring forms like furan (five membered) and pyran (six membered). In these structures, the aldehydic group in aldoses and the ketonic group in ketose change to alcoholic groups and acquire asymmetric characters.

All the monosaccharides are soluble in water, have a sweet taste, and char on heating. They give the usual chemical reactions characteristic of aldehydes or ketones. However, in the presence of dry hydrogen chloride, they exhibit a usual reaction with other hydroxy compounds such as alcohols and phenols. The products of these reactions are collectively called glycosides. Glucose gives rise to two glycosides with methyl alcohol, which are known as α-methyl glycoside and β-methyl glycoside, respectively.

By elimination of water molecules, two or more identical or different monosaccharides may join together end to end to form larger chain-like molecular structures. If only two monosaccharide molecules are joined together, the new molecule is a disaccharide. For example, if two glucose molecules are joined together, the disaccharide maltose, or malt sugar, is formed; a combination of the glucose and fructose molecule yields the disaccharide sucrose; a combination of glucose and another hexose sugar, galactose, yields disaccharide lactose, or milk sugar.

Two or more molecules of identical or different monosaccharides may join together end to end to form larger chain-like molecular structures by elimination of water molecules. For example, the disaccharides maltose, or malt sugar, sucrose, and lactose, or milk sugar, may be obtained by the combination of two glucose molecules: a glucose and a fructose molecule and a glucose and a hexose sugar or galactose molecule, respectively. The commonly known disaccharides have the molecular formula $C_{12}H_{22}O_{11}$. Disaccharides can be hydrolyzed in mild acidic solutions to give back the monosaccharides. Fructose is levorotatory, and a mixture of glucose and fructose is also levorotatory, but this is due to the strong levorotatory character of fructose. Cane sugar is dextrorotatory in nature. The mixture of sugars shows an inversion of optical rotation; this mixture is called invert sugar. Furthermore, since fructose is about twice as sweet as cane sugar, the resulting invert sugar gives extra sweetness to canned fruits and jams.

If more than two monosaccharide units are joined together, polysaccharides or polymers of simple sugars are obtained. Polypentoses $(C_5H_8O_4)_x$ are called pentosans and polyhexoses are called hexosans $(C_6H_{10}O_5)_x$. The animal polysaccharide glycogen, which occurs in muscles and in the liver, is a polymer consisting of hundreds to thousands of glucose units. The plant polysaccharide cellulose, which is present in the cell walls of the plant cells, consists of up to 2000 glucose units. Another plant polysaccharide starch, which occurs in cereal grains such as wheat and rice, consists of two polymers, amylase and amylopectin; each has about 500–1000 glucose units.

Carbohydrates function in two ways. They are either structural components of cells or function as chief biofuels to provide energy for the functioning of living systems. The monosaccharide glucose, obtained by the hydrolysis of starch or glycogen, is the chief form in which carbohydrates are transported from cell to cell in all organisms and over longer distances by sap in plants and blood in animals.

Starch is easily hydrolyzed in the presence of enzymes of human and animal digestive systems. Hydrolysis of cellulose needs enzymes of different kind. These are present in the digestive systems of grazing animals, and hence, they can use the cellulose of grass and plants as food by converting it to glucose.

3.3 THE COMPOSITION OF MARINE MICROORGANISM CARBOHYDRATES

Carbohydrates are one of the four major classes of organic compounds in living cells. They are produced during photosynthesis and are the main sources of energy for plants and animals. The term "carbohydrate" is used when referring to a saccharide or sugar and its derivatives. As discussed earlier, carbohydrates can be simple sugars or monosaccharides or double sugars or disaccharides, composed of a few sugars or oligosaccharides, or composed of many sugars or polysaccharides.

It is clear from the earlier description that there are several types of sugars, and monosaccharides change to disaccharides and polysaccharides and vice versa under a given set of conditions. The vast majority of microorganisms, including both prokaryotes (Bacteria and Archaea) and eukaryotes (phytoplankton and fungi), mostly live and grow in aggregated forms such as biofilms, which are attached to natural or man-made surfaces, as flocs and in free planktonic state (Bhaskar and Bhosle, 2005). Microorganisms are ubiquitously distributed in natural soil and aquatic environments. These microorganisms are actually embedded in a complex matrix of EPS (Prouty and Gunn, 2003). Thus, marine microorganisms may be a good source of carbohydrates. It may also be possible to obtain some unique complexes of sugar (Decho, 1990; Leppard, 1995, 1997; Santschi et al., 1998; Passow, 2002).

3.4 ANALYSIS OF CARBOHYDRATES

The stages of analysis are as follows:

1. The qualitative analysis is simple and can be performed with minimum budget. It is a basic tool to realize the basic chemistry of the analysis. It is the first stepping-stone in order to use the sophisticated, ultrasensitive, and costly instruments. Spot test analysis is a simple and inexpensive technique for the on field detection and semi-quantitative determination of the test material. It gives the firsthand

information about the presence of an analyte in a given matrix that helps in the selection of a proper analytical method with respect to its lower limit of estimation and specificity. All the characteristics of the material have been exploited in spot test analysis (Feigle and Anger, 1966). If the final product is color stuff, the nature and intensity of the color may be employed to develop a semiquantitative estimation of the test material. For this purpose, comparison of the color intensity is made with the color obtained by standard material. Thus, it also indicates the necessity of a preconcentration method to raise the concentration of the analyte up to the range of available analytical technique.

2. The removal of interfering substances from the extract—usually referred to as the cleanup procedure—often involves either chromatography or solvent partition.
3. The determination of carbohydrates together with metabolites and breakdown products in the cleaned-up extract.
4. The confirmation of the presence of the carbohydrate by using different methods after the formation and then identification of a derivative.

3.5 REACTIVE SITES OF CARBOHYDRATES

One of the most important chemical properties of monosaccharides is that they can act as mild reducing/oxidizing agents. The aldehyde group can be either oxidized to give a carboxylic group (in the presence of Cu(II), Ag(I), and Fe(III)) or reduced to the corresponding alcohol (in the presence of $NaBH_4$, KBH_4). Colorimetric methods used to analyze carbohydrates are based on these properties. For example, formation of colored products with phenolsulfuric acid, N-ethyl carbazole, anthrone, L-cysteine, L-tryptophan, or 2,4,6-tripyridyl-s-triazine (TPTZ) is due to the reducing nature of carbohydrates, and the formation of colored stuffs with 3-methyl-2-benzothiazoline hydrazone hydrochloride (MBTH) is due to its oxidizing nature. The reducing nature is also exploited in gas chromatography-flame ionization estimation (GC-FID) where carbohydrates are reduced to the corresponding alcohols prior to their derivatization. Carbohydrates act as weak acids in alkaline media, so this characteristic permits their separation by anion exchange chromatography.

3.6 DETECTION OF CARBOHYDRATES IN THE FIELD

When analyzing carbohydrates in marine organisms, the first stage is hydrolysis, which gives monosaccharides from monomers or polymers. Monosaccharides are subsequently detected by colorimetric or chromatographic methods. There are two types of hydrolysis: enzymatic and acid hydrolysis. Enzymatic hydrolysis of polymeric carbohydrates is carried out by cellular enzymes of marine bacteria. The acidic hydrolysis in marine microorganism samples generally employs hydrochloric acid (HCl) or sulfuric acid (H_2SO_4), trifluoroacetic acid (TFA) and p-toluenesulfonic acid (PTSA) are used rarely. The following tests have been reported in the literature, which may be applied for the field detection of carbohydrates.

3.6.1 Testing through Transformation to Furfural (Fegl et al., 1937)

A pinch of the sample is placed in a micro crucible or a drop of solution is taken to dryness. A drop of syrupy phosphoric acid is added, and a disk of filter paper moistened with a drop of 100% solution of aniline in 10% acetic acid is placed over the mouth of the crucible and weighted down with a watch glass. The bottom of the crucible is cautiously heated for 30–60 s with a micro flame (spattering must be avoided). A pink to red stain appears on the reagent paper. The test gives a positive response to the following carbohydrates; their concentration is given in parenthesis in γ: glucose (2.5), fructose (2.5), starch (3.0), galactose (2.5), agar-agar (5), saccharose (25), lactose (2.5), sorbose (2.5), maltose (2.5), and arabinose (2.5). Ethyl cellulose and methylcellulose, acetyl cellulose, and gum tragacanth also give a positive response at trace level (5).

Carbohydrates (di- and polysaccharides) are hydrolyzed by heating with strong acids or with oxalic acid. The monosaccharides (pentoses) are partly dehydrated on further heating to furfural or similar volatile aldehydes, such as hydroxymethylfurfural. The aldehydes are steam volatile and react with aniline to give violet Schiff bases. The hydrolysis and the dehydration are best accomplished with syrupy phosphoric acid.

3.6.2 Testing with Triphenyltetrazolium Chloride (Kuhn and Jerchel, 1941; Bondi, 1950)

A drop of test solution is mixed with two drops of 0.5% aqueous solution of triphenyltetrazolium chloride and a drop of 0.5 N sodium hydroxide in a micro test tube or micro crucible. The reaction mixture is boiled for 1–2 min. A red color or precipitate indicates a positive response for 0.2γ glucose, fructose, lactose, mannose, arabinose, and ascorbic acid.

Aldehydes do not interfere in this test, and the test gives no response for reducing agents such as hydrazine, hydroxylamine, sulfites, tartaric, and citric acids. However, ascorbic acid, which is very similar in its chemical composition to the reducing sugars, gives a positive response.

3.6.3 Testing with 5-Hydroxy-1-Tetralone (Momose and Ohkura, 1959)

A drop of the test solution is placed in a micro test tube and mixed with a drop of 0.1% alcoholic solution of 5-hydroxy-1-tetralone and 5 drops of concentrated sulfuric acid. The aliquot is kept in boiling water for 30 min, cooled, and then diluted with 1 mL water. The resulting green fluorescence is viewed in ultraviolet light. In case of concentrated sugar solution, a muddy brown color appears so it is necessary to repeat the test with the diluted solution.

The following compounds give positive response at the concentration mentioned in the parenthesis: glucose (0.2γ), mannose (0.25γ), galactose (0.3γ), fructose (0.2γ), maltose (0.3γ), lactose (0.2γ), saccharose (0.25γ), starch (0.2γ), dextrin (0.2γ), agar-agar (0.35γ), and cellulose (0.3γ). A few compounds, such as glycerol and glyceraldehydes, show fluorescence, so they interfere in the test. The negative response was shown by compounds such as pentoses, ascorbic acid, glucuronic acid, polyhydric alcohols, amino acids, aldehydes, ketones, organic acids, phenols, ethers, and proteins.

It has been claimed that the fluorescent compound produced in this case is probably due to the formation of benzo naphthalenedione. It is a faint brown compound that shows fluorescence. When the reaction mixture is diluted with much water, the brown color fades away and green fluorescence is then distinctly visible in daylight. When examined under ultraviolet light, the fluorescence becomes more evident.

3.6.4 Testing with Stannous Chloride, Sulfuric Acid, and Urea (Foulger, 1932; Zappert, 2016)

A drop of sugar solution is treated with 6–20 drops of the reagent solution in a micro crucible and then heated over a micro flame. After cooling, a blue color appears if keto-hexoses are present. A less intense red color develops due to aldohexoses only after longer boiling.

The test gives a positive response for fructose (8γ), insulin (10γ), saccharose (15γ), and sorbose (8γ). The reagent was prepared by heating 8 g urea and 0.2 g stannous chloride with 10 mL of 40% sulfuric acid. The mechanism of this color formation is still unknown.

3.7 ISOLATION AND PURIFICATION METHODS OF EXTRACELLULAR POLYMERIC SUBSTANCES

Among several interactions, van der Waals forces, electrostatic interactions, hydrogen bonds, hydrophobic interactions, and covalent bonds such as disulfide bonds in glycoprotein are the main forces between EPSs and the cell surface (Christensen and Characklis, 1990; Wingender et al., 1999; Xu, 2007). However, the dominant forces might be very different from one EPS matrix to another. Furthermore, in studies with different chemically identifiable target components, for example, extracellular proteins versus extracellular polysaccharides, the extraction method could be quite different even for the same EPS matrix. Therefore, many methods have been proposed for extraction of capsular polymeric substances from the cell. There are no universal methods available, and so comparative experiments of various methods are needed. Therefore, in most cases, physical methods (ultracentrifugation, ultrasonic stirring, autoclaving, cation exchange chromatography) and chemical methods (addition of mild extractants such as tap water and sodium chloride, strong extractants such as sodium hydroxide and ammonium hydroxide, EDTA, acidic reagents such as sulfuric acid and hydrochloric acid, and aldehyde) are combined together to obtain the maximum yield. These methods may be subdivided in two groups: commonly used affordable methods and ultrasensitive, sophisticated, and costly methods.

3.7.1 Common Methods of Extraction

3.7.1.1 Dialysis or Diafiltration

Simple and affordable isolation and purification methods such as freeze concentration, lyophilization, evaporation, distillation, reverse osmosis, dialysis, electrophoresis, ultrafiltration, liquid–liquid extraction, solid–liquid extraction, and precipitation have been reported in the literature (Rathore and Khan, 2000). The commonly known methods that have been used for the analysis of carbohydrates in marine microorganisms are summarized in the following paragraphs:

Dialysis is a method used to separate molecules through a semipermeable membrane. The concentration gradient of components across the membrane drives the separation. Dialysis is used for the removal of excess low-molecular-weight solutes. Many scientists have used this technique for the isolation and cleanup of extracellular polymeric substances

(Cerning et al., 1988; Marshall et al., 1995; van Kranenburg et al., 1997; Gehrke et al., 1998; Beech et al., 1999; Bergmaier et al., 2001; Chester, 2003; Veningelgem et al., 2004; Hung et al., 2005; Alvarado et al., 2006). All the contaminants small enough to diffuse through the membrane pores are removed.

3.7.1.2 Precipitation

It is known that materials such as humic acids are soluble in bases but insoluble in alcohols. Therefore, materials like humic acids may be precipitated by acidification with carboxylic acids such as glacial acetic acid in the presence of alcohols. In the area of marine microorganisms, carbohydrates, ethanol, methanol, acetone, and so on have been used to precipitate out small molecules like ions, but proteins and other macro-molecules could also be coprecipitated (Van Geel-Schutten et al., 1999; Petry et al., 2000; Torino et al., 2000b; Dal Bello et al., 2001; Degeest et al., 2001; Rimada and Abraham, 2001; Ricciardi et al., 2002; Hung et al., 2005). Trichloroacetic acid has also been used in purification of extracellular polymeric substances. But in this case, amino acids, peptides, proteins, and some polysaccharides might also be coprecipi-tated (Cerning et al., 1944; Gaecia-Garibay and Marshall, 1991; Grobben et al., 1995; Van Marle and Zoom, 1995; Dupont et al., 2000; Frengova et al., 2000; Knoshaug et al., 2000; Ruas-Madeodo et al., 2002; Harding et al., 2003; Ruas-Madiedo and de los, 2005).

3.7.1.3 Enzyme Digestion

Enzyme digestion is a routine method of separation/cleanup in molecular biology. Till now, researchers use enzyme digestion for cloning of genome and DNA. Enzyme immo-bilized silica resin column chromatography has been used to purify DNA. Enzyme diges-tion method is used to remove proteins and nucleic acids by using protease and nuclease, respectively (Cerning et al., 1986, 1992; Abbad-Andaloussi et al., 1995; Bouzar et al., 1996, 1997; Mozzi et al., 1996; Rathore and Khan, 2000; Torino et al., 2000a, 2001; Hung et al., 2005).

3.7.1.4 Size Exclusion Chromatography

Size exclusion chromatography is a popular method to separate biomolecules based on their size. Primarily, it is applied to the separation of biopolymers. As this technique provides separations of the molecules of different size, that is, molecular weight, all the contaminants with the molecular weights different from target compounds are removed (Chen et al., 1996; Hwang et al., 2003; da Silva et al., 2005; Yang et al., 2005).

3.7.1.5 Anion Exchange Chromatography

Ion exchange chromatography involves the separation of ionizable molecules based on their total charge. This technique enables the separation of similar types of molecules on the basis of their charge.

   Therefore, this technique has been used to remove any macromolecules with surface net charges different from polysaccharides (Wu et al., 2005; Denkhaus et al., 2007; Yang et al., 2016).

3.7.1.6 Single-Dimension Gel Electrophoresis

This technique separates molecules on the basis of their charge to size ratio. Charged molecules move under the influence of an electric field at a constant pH (buffer system).

Single-dimension gel electrophoresis, that is, isoelectric focusing, has been utilized to remove all the macromolecules with isoelectric point different from polysaccharides (Santschi et al., 2003; Alvarado et al., 2006; Ray, 2006).

3.7.1.7 Two-Dimensional SDS-PAGE

It is a very common method for separating proteins by electrophoresis. It uses a discontinuous polyacrylamide as a support medium and sodium dodecyl sulfate to denature the proteins. Sodium dodecyl sulfate (SDS) is a detergent that dissociates and unfolds oligomeric proteins into its subunits. The method is called sodium dodecyl sulfate polyacrylamide gel electrophoresis (SDS-PAGE). This method has been used to remove any macromolecules with surface net charges and molecular masses different from polysaccharides (Quigley et al., 2002; Alvarado et al., 2006; Ray, 2006).

3.7.2 Ultrasensitive, Sophisticated, and Costly Methods

Traditional extraction techniques such as Soxhlet, solid–liquid extraction (SLE), and liquid–liquid extraction (LLE) require large volumes of solvent and are time-consuming. They often produce low yields of bioactives and their selectivity is poor. These techniques are usually not automated and their reproducibility can therefore be compromised. It is also important to consider how functional ingredients are obtained from new matrices such as micro- and macroalgae. Thus, there is a need to develop appropriate, selective, affordable (cost-effective), and environmentally friendly extraction procedures that can provide the legal requirements regarding the use of food-grade solvents and processes. Therefore, several modern extraction techniques have been developed, and they may be divided in two subgroups (Santschi et al., 2003), that is, fluid-phase partitioning methods such as single-drop and liquid microextraction, supercritical and pressurized liquid extraction, accelerated solvent extraction, pressurized hot water extraction, ultrasound-assisted extraction, microwave-assisted extraction, and simple fluid-phase partitioning extraction, and sorptive and membrane-based methods such as solid-phase microextraction, sorptive-phase developments, and hollow-fiber membrane extraction. These techniques alone of their combination may provide an effective alternative to the problems encountered with the use of traditional extraction procedures. The selection of the procedure to be employed is governed by the type of analyte and the nature of the sample under consideration. The methods used for extraction and cleanup of carbohydrates in marine microorganisms are discussed in the following paragraphs.

3.7.2.1 Supercritical Fluid Extraction

Supercritical fluid extraction (SFE) is based on the use of solvents at temperatures and pressures above their critical points. It is a good alternative to conventional methods such as SLE and LLE, which utilize a large volume of hazardous chemicals such as chlorinated solvents. SFE was employed previously to extract a wide variety of interesting compounds from different food stuffs including algae (Herrrero et al., 2005; Mendiola et al., 2007). The important features of SFE are negligible use of toxic solvents, cost-effective, easily attainable critical conditions (30.9C and 73.8 bars), and eco-friendly nature of CO_2, that is, generally recognized as safe (GRAS) for use in food industry. This technique has been successfully applied for the extraction of bioactives including carbohydrates from marine resources (Ibanez et al., 2012).

3.7.2.2 Pressurized Liquid Extraction

This technique is known by different names in different places (Nieto et al., 2010), that is, pressurized fluid extraction, enhanced solvent extraction, high-pressure solvent extraction, or ASE. It was introduced in 1966 (Richter et al., 1996). In PLE, pressure is applied in order to use the solvents at a temperature above their normal boiling point. ASE may be considered as a new version of Soxhlet apparatus, which operates at a higher pressure and temperature. Some other techniques, namely, pressurized hot water extraction (PHWE), subcritical water extraction (SWE), near-critical fluid extraction, and enhanced fluid extraction, have been found to be most promising in bioactives including carbohydrate extraction from different raw materials (marine microorganisms) (Herrero et al., 2006; Rathore, 2010). These techniques require a small amount of solvent (10–50 mL) and a small period of time (20 min), while traditional extraction procedures require more solvent and time (300 mL and 10–48 h). Both the solubility as the rate of mass transfer increase with increasing temperature; in turn, there is increase in analytical solubility. The elevated temperature also decreases the viscosity and the surface tension of the solvent. It improves the extraction rate. Thus, these techniques have been recognized as green extraction methods. These procedures have been applied to extract several bioactives including carbohydrates in marine microorganisms (Isaac et al., 2005; Spiric et al., 2010).

3.7.2.3 Pressurized Hot Water Extraction

This is a specialized form of PLE in which water is used as an extracting solvent. It is designated as pressurized hot water extraction (PHWE), subcritical water extraction, pressurized low-polarity water extraction (PLPW), or superheated water extraction (SHWE). Water is a well-known versatile eco-friendly solvent, and it is used to isolate functional ingredients from different raw materials including plants and food waste. This technique is based on the application of liquid water at temperature above its atmospheric boiling point that is maintained by applying pressure. Physical and chemical properties of water are drastically changed under these conditions, for example, dielectric constant decreases from around 80 at 25°C to around 33 at 2000°C (very close to methanol). Other parameters such as viscosity and surface tension both are reduced with increasing temperature, while diffusivity is increased as a result efficiency and speed of extraction is improved. In addition, solubility of different compounds present in water is also modified and their selectivity is enhanced. As a whole, there is good control of temperature, pressure, extraction time, flow rate, selectivity, and so on in this technique. Several applications of this technique in the area of food stuffs, drugs, and medicinal plants are published in the literature (Yoshida et al., 1999; Ong et al., 2006; Wang and Weller, 2006; Gol Mohammad et al., 2008; Teo et al., 2010; Turner and Ibanez, 2011). Applications of this technique to marine by-products have been published in the form of many research papers (Yoshida et al., 1999, 2003; Kang et al., 2001).

3.7.2.4 New Approaches

Ultrasound-assisted extraction (UAE) and Microwave-assisted extraction (MAE) are recently developed techniques used for extraction with minimum health hazards/environmental pollution. They are rapid and cost-effective as there is a minimum amount of eco-friendly solvents required. UAE uses acoustic cavitation to cause disruption of cell walls, reduction of particle size, and increase in contact period between the solvent and the target compound. The microwave radiation used in MAE causes motion of polar molecules and rotation of dipoles to heat solvent and to promote transfer of target compounds from the sample into the solution (Ying et al., 2011). Both the procedures are claimed to be versatile as they provide the possibility of using several solvents possessing different polarity as well as both can couple extraction and reaction simultaneously. Both the techniques have not been installed in many laboratories so far, so they have not been fully exploited (Batista et al., 2001). The earlier description reflects the versatility of pressurized solvents due to their largely different physicochemical properties—density, diffusibility, viscosity, dielectric constant, and so on—and control of these parameters, which results in controlling solvating power and selectivity of the solvent.

3.8 QUANTIFICATION OF MARINE MICROORGANISM CARBOHYDRATES

In quantification, the first step is the hydrolysis of the marine matrix to obtain simple sugar/sugars. Acids of different concentration have been used for different time periods and temperature to hydrolyze the marine microorganisms' samples in order to have monosaccharide for estimation. Some pertinent examples are 1.8 N HCl for 3.5 h and 100°C, 0.5 M H_2SO_4 for 4 h at 100°C, 4 M HCl for 3 h at 110°C, 1.2 M H_2SO_4 for 3 h at 100°C, 1 M H_2SO_4 for 4 h at 90°C, 0.5 M HCl for 1 h at 100°C, 2 M HCl for 3.5 h at 100°C, 0.5 N Trifluoroacetic acid for 2 h at 135°C, 0.75 M Para toluene sulfonic acid (ptsa) + glycerin for 4 h at 100°C, extracellular enzyme of marine bacteria, and so on. The second step is the conversion of the sugar in color stuff that can be quantified by visual comparison or colorimetry or spectrophotometry. The simple sugar can be quantified by thin-layer chromatography (TLC), high-performance thin-layer chromatography (HPTLC), column chromatography (LC), high-performance chromatography (HPLC), electrochemical methods (electrophoresis or capillary electrophoresis), or radiochemical methods. The simple sugar can also be derivatized to obtain a volatile product that can be quantified by gas chromatography (GC) or gas chromatography-mass spectrometry or mass spectrometry (GC-MS) or capillary gas chromatography (CGC). The methods reported in the literature are summarized in the following paragraphs.

3.8.1 Colorimetric/Spectrophotometric Method

The colorimetric method is applicable in the milligram concentration range so it gives semiquantification. It is very simple and inexpensive and can be performed on field analysis. Spectrophotometry is also a simple, inexpensive, and routinely used method. It gives results in microgram rang. However, it does not achieve the sensitivity and selectivity or specificity of chromatographic methods. Dubois et al. (Dobois et al., 1956) have used phenol–sulfuric acid as a coloring reagent for the spectrophotometric quantification of total carbohydrates (TCHO) in marine microorganisms. The phenol–sulfuric acid (PSA) method is also known as the Dubois method. In this method, sugars are dehydrated in the presence of concentrated sulfuric acid at an elevated temperature to yield furfurals from pentoses or hydroxymethylfurfural from hexoses. These products condense with phenol to give yellow-orange stuffs that absorb at 480–490 nm. The intensity of the color is proportional to the sugar concentration at a fixed concentration of phenol. The method possesses good precision (<20% at 50 µM) and detection limits (25–50 µM). This method has also been claimed to give good results for methyl derivatives including uronic acids, some oligosaccharide (raffinose and sucrose) and polysaccharides (dextran and starch).

It is a simple and rapid method; its sensitivity depends on the matrix and it has different molar sensitivities for different groups of sugars. The method is unable to distinguish between mono- and polysaccharide. It is used for the estimation of sugars in particulate or the sedimentary organic materials and is rarely used for seawater samples.

Some analysts have also used MBTH (3-methyl-2-benzothiazolinone hydrazone hydrochloride) as a color-forming agent with sugars (Burney and Sieburth, 1977; Johnson and Sieburth, 1977; Johnson et al., 1981). The color formation is based upon three reactions: (1) reduction of monosaccharides to corresponding alditols by potassium borohydride, (2) reaction of the alditols so produced with periodic acid to give formaldehyde, and (3) reaction of the latter compound with chromogen and MBTH, to produce a blue complex that absorbs at 635 nm. This method is specific for sugars and applicable in seawater. It has a low limit of detection (420–500 µM) and precision (<10% at 1 µM level). This method has been applied for the estimation of major classes of sugars (pentoses, hexoses, and deoxy sugars), amino sugars, uronic acids, and some disaccharides (maltose). However, it is a laborious and lengthy method as it involves three reactions.

TPTZ has been used successfully for the spectrophotometric determination of TCHO in marine microorganisms (Hung et al., 2001; Witter and Luther III, 2002). The TPTZ method is the recently developed method for the spectrophotometric determination of carbohydrates in marine organisms. Its color formation depends on the oxidation of polysaccharides by $(K_3(Fe(CN)_6I))$ in alkaline medium (Avigad, 1968). The complex so formed reacts with TPTZ to produce a violet complex that absorbs at 595 nm. This method, like the MBTH method, is less sensitive and possesses low detection limits (0.4 µg) and precision 10% (580 nM). However, it is simple and rapid like the PSA method. This method is applicable to estimate both mono- and polysaccharides in saline water. Anthrone–sulfuric acid is used as a coloring reagent by many scientists (Morris, 1948; Levander et al., 2001; Rimada and Abraham, 2001) for the determination of TCHO.

Some other chromogens such as ruthenium red adsorption (Figueroa and Silverstein, 1989) and Alcian Blue strain (Passow and Alldredge, 1995) for APS and meta hydroxyphenyl (Filisetti-Cozzi and Carpita, 1991; Hung and Santschi, 2001) for uronic acids (APS) have been used for the spectrophotometric quantification.

3.8.2 Chromatography

3.8.2.1 Paper Chromatography

The most simple, low-cost, and rapid chromatographic technique, paper chromatography (PC), was the first method to be routinely applied for the separation of sugars in sediment hydrolysates. It does not give reproducible results so it has been replaced by thin-layer chromatography.

3.8.2.2 Thin-Layer Chromatography

TLC is performed on an adsorbent thin layer that is coated on strong and firm support such as glass, metal, and plastic plates. The coating materials that are generally used are silica gel or cellulose (Plunket, 1957) including many other stable, insoluble, porous, solid adsorbents. This method has been employed for the separation of monosaccharides and their derivatives by the use of different solvents (Ghebregzabher et al., 1976). The position, shape, size, and density/intensity of the sugar spot on the plate are the parameters which are used for the qualitative or semiquantitative analysis of the analyte. The demarcated sugar spot can be scratched out, and the sugar can be quantified by spectrophotometric method using a chromogen such as tetrazolium blue (Mopper and Degens, 1972). TLC was mainly applied in freshwater, seawater, POM (particulate organic matter), and sediments (Degen et al., 1865; Artem'ev, 1874; Handa and Tominaga, 1969). TLC has been found to be time-consuming, and it requires a larger amount of the test material. The separation of oligosaccharides is not possible (Olechno et al., 1987) on TLC plates. It may be only used as a pilot technique to column chromatography.

3.8.2.3 Low-Pressure Liquid Chromatography

Low-pressure liquid chromatography (LC) and high-pressure liquid chromatography or high-performance liquid chromatography (HPLC) both have been used successfully in analyzing saccharides. In the area of LC, the first approach was made by Larsson and Samuelson (1967). He studied anion exchange resins column with ethanol–water as eluant. Detection was made by chromogens such as orcinol–sulfuric acid (Josefesson, 1970), tetrazolium blue (Mopper and Degens, 1972), or Cu-bicinchoninate (Mopper and Gindler, 1973; Mopper, 1978a). This method detects digitoxose, 2-deoxy ribose, 2-deoxy ribose galactose, 3-O-methyl glucose, 6-deoxy glucose, fucose, rhamnose, arabinose, galactose, glucose, mannose, xylose, fructose, ribose, lyxose, tagatose, sorbose, and glucose (Mopper, 1978a). Analysis of sediments and dissolved organic matter (DOM) of seawater shows that the major sugars are fucose, rhamnose, arabinose, galactose, glucose, mannose, xylose, fructose, and ribose. Glucose and sorbose were found in trace.

The studies of borate complexes on strong anion exchange resins with Cu-bicinchoninate (Mopper, 1978c) or ethylenediamine (Mopper et al., 1980) detection was the first procedure for the analysis of environmental samples including seawater. This technique has been used on anion exchange resins column in borate medium to detect the following sugars: 2-deoxy ribose, cellobiose, maltose, lactose, fucose, rhamnose, arabinose, galactose, glucose, mannose, xylose, fructose, ribose, gentiobiose, and melibiose (Mopper, 1978c). This technique has been used to analyze sinking or suspended particulate organic matter (POM), and major sugars were found to be fucose, rhamnose, arabinose, galactose, glucose, mannose, xylose, and ribose (Mopper et al., 1980; Ittekkot et al., 1982, 1984a). However, the analysis of DOM in seawater indicated the presence of fructose (Ittekkot et al., 1984b).

Anion exchange resins column in acetate medium detects the presence of mannuronic, glucuronic, guluronic, galacturonic, 4-O-methyl glucuronic, and cellobiuronic acids (Mopper, 1977, 1978b).

3.8.2.4 High-Pressure Liquid Column Chromatography

High-pressure liquid chromatography (HPLC) is also known as high-performance liquid column chromatography. It is fast and selective; it requires a minute quantity of the analyte and gives better resolution. But it is costly and sophisticated, so it has not been installed in many laboratories of third-world countries so far. In marine microorganisms' carbohydrate analysis, HPLC has been applied either on reversed-phase columns using derivatives of 5-diamino-naphthalene-sulphonydrazine (DNS) and p-amino benzoic acid (p-AMBA) derivatives or on an anion exchange column.

3.8.2.4.1 High-Pressure Anion Exchange Chromatography–Pulse Amperometric Detection
HPAEC hyphenated with pulse amperometric detection (PAD) is permitting the detection of carbohydrates including those without reducing group/groups with high sensitivity (<10 pmol) without pre- or postcolumn derivatization (Rocklin and Pohl, 1983;

Olechno et al., 1987; Johnson and LaCourse, 1990; Lee, 1990). The first application of HPAEC-PAD was developed in the 1990s (Mopper et al., 1992; Jorgensen and Jensen, 1994; Rich et al., 1996; Borch and Kirchman, 1997; Skoog and Benner, 1997). PAD applies a triple sequence of potential to a gold electrode, and it is successfully used to determine at nM levels (Rocklin and Pohl, 1983; Mopper et al., 1992). In marine micro-organisms' carbohydrates analysis, this method has been mainly used to analyze sugars in DOM (dissolved organic matter) and UDOM (ultrafiltered dissolved organic matter) (Jorgensen and Jensen, 1994; Rich et al., 1996, 1997; Borch and Kirchman, 1997; Gremm and Kaplan, 1997; Skoog and Benner, 1997; Amon et al., 2001; Kirchman et al., 2001; Amon and Benner, 2003; Benner and Kaiser, 2003). POM and sediments sample have also sugars being analyzed by using this method (Buscail et al., 1995; Kerherve et al., 1995, 1999, 2002; Skoog and Benner, 1997; Panagiotopoulos and Sempere, 2005b). This method has also applied for analyzing sugars present in POM (particulate organic matter) and in sediments.

3.8.2.4.2 Reversed-Phase High-Pressure Liquid Chromatography This device consists of a precolumn for derivatization of sugars with DNS or p-AMBA, octadecylsilyl or a C18 column, and a suitable detector. Derivatization decreases the polarity of sugars, shortens the time of separation, improves the elution, and minimizes the interferences due to salts and other contaminants. Several reagents have tried for sugar derivatization (Honda, 1996). DNS derivatization has been used to analyze seawater (Senior et al., 1985), sinking and suspended POM (Compiano et al., 1993; Hicks et al., 1994), and sediment samples (Hicks et al., 1994). The elution has been performed either by gradient procedure and fluorimetric detection at excitation 360–380 nm and emission 540 nm (Mopper and Johnson, 1983) or by isocratic mode and UV detection at 230 nm (Senior et al., 1985). The precision and detection limits are <10% at the μM level and 200–500 nM, respectively. The corrected version is that the precision of the method under study is <10% at μM level with a detection limit of 200–500 nM. This method analyses neutral sugars and some reducing oligosaccharides, namely, cellobiose, lactose, maltose, and gentiobiose. Carbonyl compounds such as simple aldehydes and ketones or ketosteroids interfere in the procedure. Despite some drawbacks, this method has been found to be simple and applicable to reducing sugar analysis with minimum pretreatment of the test material.

The p-AMBA derivatization method has also been used for the determination of sugars in environmental samples (Meyer et al., 2001). The detection was made with photometric detection at 303 nm or fluorimetric detection at excitation 313 nm and emission at 358 nm. There should be proper control and optimization of pH of the eluant, temperature of derivatization, and eluant composition. This method has been used for the analysis of neutral sugars, namely, xylose, arabinose, galactose, glucose, and mannose and some amino sugars including *N*-acetyl glucosamine.

3.8.2.5 Gas Chromatography-Flame Ionization Detection

Being nonvolatile, carbohydrates may not be determined directly by GC. Efforts have been made to convert sugars in volatile derivatives by using trimethylsilyl ethers (Li and Andrews, 1986; Hernes et al., 1996), trifluoroacetate esters (Eklund et al., 1977), and acetyl derivatives (Klok et al., 1984; Sakugawa and Handa, 1985; Bhosle et al., 1992). The precision of the GC-FID method ranges from 10% to 20% depending on sugar concentration (MacCarthy et al., 1996; D'Souza and Bhosle, 2001; Oggier et al., 2001). The limit of detection ranges from 100 to 1150 nM (Cowie and Hedges, 1984a).

This technique has been used with narrow (capillary) or broad column for analysis of most common saccharides, namely, fucose, rhamnose, arabinose, galactose, glucose, mannose, and ribose (Cowie and Hedges, 1984a). It is also suitable for the determination of uronic acids such as glucuronic acid, mannuronic acid, guluronic acid, and galacturonic acids (Walter and Hedges, 1988). Attempts have also been made for the simultaneous determination of neutral and acidic sugars (Bergamaschi et al., 1999). This technique suffers two main drawbacks: First, sugars in solution may exist in five different forms (one acyclic form and two anomers for each of the five- and six-membered ring forms). This multiplicity leads to complex peaks for each sugar that is difficult to dissolve. Second, some of the derivatives are less stable (TFA derivatives) and may be lost in the chromatographic column (Eklund et al., 1977). The technique is also less sensitive than HPLC. So, it has limited use especially for the analysis of dissolved free monosaccharides.

3.8.3 Mass Spectrometry

The progress is slow in carbohydrate (glycan) analysis by MS; which may be due to the high cost and sophistication of this technique. Being very sensitive and informative, MS is the only technique that provides full details of structure. The characterization of glycans relies upon obtaining the minor details of structure. Subtle differences due to isomerism or chirality can produce molecules with very different biological activities, making complete structural analysis even more demanding. In fact, carbohydrates are central players in a number of important biological processes including cell signaling, cell adhesion, and the regulation of biochemical pathways. They occur in nature as heterogeneous mixtures, often of high complexity.

The identity (species) and the level (concentration) of glycan expression (signal) within animal cells and tissues are known to affect physiological and pathological processes and to correlate with the health of an individual (Dube and Bertozzi, 2005; Ohtsubo and Marth, 2006). The level of expression of these vital molecules varies between individual species, and they are known to change during the development and progression of many diseases (Prescher and Bertozzi, 2006; Nairn et al., 2008). The quantification of glycans in cells and tissues is a very important aspect in marine microorganisms.

Therefore, there is genuine interest in improving mass spectrometry methodologies and technologies in this decade. The recent developments in this area have benefited carbohydrate analysis. The developments include approaches for improved ionization, new and improved methods of ion activation, advances in chromatographic separation of carbohydrates, hybridization of ion mobility, mass spectrometry, and better software data collection and interpretation. A critical review about recent developments in the application of mass spectrometry to the analysis of carbohydrates is published by Kailemia et al. (2014).

3.8.4 Enzymatic Methods

Being based on a specific reaction, the enzymatic techniques are highly selective and very sensitive (10–30 nM) (Olechno et al., 1987). Enzymatic methods suffer badly due to the presence of metal salts and other contaminants, which deactivate the enzyme (Dawson and Liebezeit, 1981). Efforts have been made to quantify glucose in seawater (Hicks and Carey, 1968; Hanson and Snyder, 1979). But results obtained are poor and attempts have not been made for other saccharides or total saccharide (Guilbaut, 1984).

3.8.5 Other Techniques

Isoelectric focusing electrophoresis (IEF) is an electrophoretic method that separates molecules based on their isoelectric point (pi). After proper development for about 17–18 h, the strips were carefully cut evenly into 11 fractions, and the same fraction (portion) in every strip was pooled and extracted in 10% sodium dodecyl sulfate (SDS) for 24 h, and sugar was determined using a suitable method. It is a time-consuming technique and may be used to separate sugars present in a complex mixture of marine microorganism (Xu, 2007). 234[Th] (IV) and 240 Pu activity analysis has also been applied to analyze carbohydrates in marine microorganisms (Quigley et al., 2001; Alvarado et al., 2006). It is a most sensitive, costly, and injurious technique and it has restricted use in carbohydrate analysis.

REFERENCES

Abbad-Andaloussi, S., Talbaouiu, H., Marczak, R., and Bonaly, R. Isolation and acterization of exocellular polysaccharides produced by *Bifidobacterium longum*. *Appl. Microbiol. Biotechnol.*, 43 (1995): 995–1000.

Alvarado, N.A., Hung, C.C., and Santschi, P.H. Binding of thorium (IV) to carboxylate, phosphate and sulphate functional groups from marine exopolymeric substances (EPS). *Mar. Chem.*, 100 (2006): 337–352.

Amon, R.M.W. and Benner, R. Combined neutral sugar as indicators of the diagenetic state of dissolved organic matter in Arctic Ocean. *Deep-Sea Res.*, I 50 (2003): 151–169.

Amon, R.M.W., Fitznar, H.P., and Benner, R. Linkages among the bioreactvity, chemical composition and diagenetic state of marine dissolved organic matter. *Limnol. Oceanogr.*, 46 (2001): 287–297.

Artem'ev. Comparison of the composition of carbohydrates in phytoplankton, suspended matter and bottom sediments of the ocean. *Oceanology*, 14 (1874): 832–835.

Avigad, G. A modified procedure for the colorimetric, ultramicro determination of reducing sugars with alkaline ferricyanide reagent. *Carbohyd. Res.*, 7 (1968): 94–97.

Batista, A., Vetter, W., and Luckas, B. Use of focused open vessel microwave-assisted extraction as prelude for the determination of the fatty acid profile of fish—A comparison with results obtained after liquid-liquid extraction according to Bligh and Dyer. *Eur. Food Res. Technol.*, 212 (2001): 377–384.

Beech, I., Hanjagsit, L., Kalaji, M., Neal, A.L., and Zinkevich, V. Chemical and structural characterization of exopolymers produces by Pseudomonas sp. NCIMB.2021 in continuous culture. *Microbiology*, 145 (1999): 1491–1497.

Benner, R. and Kaiser, K. Abundance of amino sugars and peptidoglycan in marine particulate and dissolved organic matter. *Limnol. Oceanogr.*, 48 (2003): 118–128.

Bergamaschi, B.A., Walters, J.S., and Hedges, J.I. Distribution of uronic acids and O-methyl sugars in sinking and sedimentary particles in two coastal marine environments. *Geochim. Cosmochim. Acta*, 63 (1999): 413–425.

Bergmaier, D., Lacroix, C., Guadalupe Macedo, M., and Champagne, C.P. New method for exopolysaccharide determination in culture broth using stirred ultrafiltration cells. *Appl. Microbiol. Biotechnol.*, 57 (2001): 401–406.

Bernan, V.S., Greenstein, M., and Maiese, W.M. Marine microorganisms as a source of new natural products. *Adv. in Appl. Microbiol.*, 43 (1997): 57–90.

Bhaskar, P.V. and Bhosle, N.B. Microbial extracellular polymeric substances in marine biogeochemical process. *Curr. Sci.*, 88(1) (2005): 45–53.

Bhosle, N.B., Sankran, P.D. and Wagh, A.B. Monosaccharides composition of suspended particles from the Bay of Bengal. *Oceanol. Acta*, 15 (1992): 27286.

Bondi, A. (Rehovoth). Unpublished studies; see also Mattson, A.N. and Tenen, C.O., *Anal. Chem.*, 22 (1950): 183; Mattson, A.N. and Tenen, C.O., *Science*, 106 (1947), 294.

Borch, N.H. and Kirchman, D.L. Concentration and composition of dissolved combined neutral sugars (polysaccharide) in seawater determined by HPLC-PAD. *Mar. Chem.*, 57 (1997): 85–95.

Bouzar, F., Cerning, J., and Desmazeaud, M. Exopolysaccharide production in milk by *Lactobacillus delbrueckii* spp. Bulgaricus CNRZ 1187 and two colonial variants. *J. Dairy Sci.*, 79 (1996): 205–211.

Bouzar, F., Cerning, J., and Desmazeaud, M. Exopolysaccharide production and texture-promoting abilities of mixed-strain starter cultures in yogurt production. *J. Dairy Sci.*, 80 (1997): 2310–2317.

Burney, C.M. and Sieburth, J.McN. Dissolved carbohydrate in seawater. 2. A spectrophotometric procedure for total carbohydrate analysis and polysaccharide estimation. *Mar. Chem.*, 5 (1977): 15–28.

Buscail, R., Pochlington, R., and Germain, C. Seasonal variability of the organic matter in a sedimentary coastal environment: Sources, degradation and accumulation (continental shelf of the Gulf of Lions-Northwestern Mediterranean Sea). *Cont. Shelf. Res.*, 15 (1995): 843–869.

Cerning, J., Bouillanne, C., Demazeaud, M.J., and Landon, M. Isolation and characterization of exocellular polysaccharide produced by *Lactobacillus bulgaricus*. *Biotechnol. Lett.*, 8 (1986): 625–628.

Cerning, J., Bouillane, C., Desmazeaud, M.J., and Landon, M. Exocellular polysaccharide production by *Streptococcus thermophilus*. *Biotechnol. Lett.*, 10 (1988): 255–260.

Cerning, J., Bouillanne, C., Landon, M., and Desmazeaud, M.J. Isolation and characterization of exopolysaccharides from slime-forming mesophilic lactic acid bacteria. *J. Dairy Sci.*, 75 (1992): 692–699.

Cerning, J., Renard, C.M.G.C., Thibault, J.F., Bouillanne, C., London, M., Desmazeaud, M., and Topisirovic, L. Carbon source requirements for exopolysaccharide production by *Lactobacillus casei* CG11 and partial structure analysis of the polymer. *Appl. Environ. Micribiol.*, 60 (1944): 3914–3919.

Chen, L., Wilson, R.H., and McCann, M.C. Investigation of macromolecule orientation in dry and hydrated walls of single anion epidermal cells by FTIR microspectroscopy. *J. Mol. Struct.*, 408 (1996): 257–260.

Chester, R. *Marine Geochemistry*, 2nd edn. Blackwell, Oxford, UK, 2003.

Christensen, B.E. and Characklis, W.G. Physical and chemical properties of biofilms. In: *Biofilms*. Characklis, W.G. and Marshal, K.C., Eds. John Wiley & Sons, Toronto, Ontario, Canada, 1990, pp. 69–130.

Compiano, A.M., Romano, J.C., Carabetian, F., Laborde, P., and de la Giraudiere, I. Monosaccharide composition of particulate hydrolysable sugar fraction in surface microlayers from brackish and marine waters. *Mar. Chem.*, 42 (1993): 237–251.

Cowie, G.L. and Hedges, J.I. Determination of neutral sugars in plankton, sediments and wood capillary gas chromatography of equilibrated isomeric mixtures. *Anal. Chem.*, 56 (1984a): 487–504.

da Silva, M.L.C, Izeli, N.L., Martinez, P.F., Silva, I.R., Constatino, C.J.L., Cardoso, M.S., Barbosa, A.M., Dekker, R.F.H., da Silva, G.V.L.J. Purification and structural characterization of (1→3;1→6)-β-D-glucans (botryosphaerans) from *Botryosphaeria rhodina* grown on sucrose and fructose as carbon sources: A comparative study. *Carbohyd. Polym.*, 61 (2005): 10–17.

Dal Bello, F.D., Walter, J., Hertel, C., and Hammes, W.P. In vitro study of prebiotic properties of levan-type exopolysaccharides from lactobacilli and non-digestible carbohydrates using denaturing gradient gel electrophoresis. *Syst. Appl. Micribiol.*, 24 (2001): 232–237.

Dawson, R. and Liebezeit, G. The analytical methods for characterization of organics in seawater. In: *Marine Organic Chemistry: Evolution, Composition, Interactions and Chemistry of Organic Matter in Seawater.* Dousman, E.K. and Dawson, R., Eds. Oceanography Series No.31, Elsevier Scientific, Amsterdam, the Netherlands, 1981, pp. 445–496.

Decho, A.W. Microbial exopolymer secretions in ocean environments: Their role(s) in food webs and marine processes. *Oceanogr. Mar. Biol. Annu. Rev.*, 28 (1990): 73–153.

Degeest, B., Janssens, B., and De Vust, L. Exopolysaccharide (EPS) biosynthesis by *Lactobacillus sakei* 0–1: Production kinetics, enzyme activities and EPS yields. *J. Appl. Microbiol.*, 91 (2001): 470–577.

Degen, E.T., Reuter, J.H., and Shaw, K.N.F. Biochemical compounds in offshore California sediments and seawaters. *Geochim. Cosmochim. Acta*, 28 (1865): 45–66.

Denkhaus, E., Meisen, S., Telgheder, U., and Wingender, J. Chemical and physical methods for characterization of biofilms. *Microchim. Acta*, 158 (2007): 1–27.

Dobois, M., Gills, K.A., Hamilton, J.K., Rebers, P.A., and Smith, F. Colorimetric method for determination of sugars and related substances. *Anal. Chem.*, 28 (1956): 350–356.

D'Souza, F. and Bhosle, N.B. Variation in the composition of carbohydrates in the Dona Paula Bay (West India) during May-June 1998. *Oceanol. Acta*, 24 (2001): 221–237.

Dube, D.H. and Bertozzi, C.R. Glycans in cancer and inflammation—Potential for therapeutics and diagnostics. *Nat. Rev. Drug Discov.*, 4 (2005): 477–488.

Dupont, I., Roy, D., and Lapointe, G. Comparison of exopolysaccharide production by strains of *Lactobacillus rhamnosus* and *Lactobacillus paracasei* grown in chemically defined medium and milk. *J. Ind. Microbiol. Biotechnol.*, 24 (2000): 251–255.

Eklund, G., Josefesson, B., and Roos, C. Gas-liquid chromatography of monosaccharides at the pictogram level using glass capillary columns, trifluoacetyl derivatization and electron-capture detection. *J. Chromatogr.*, 142 (1977): 575–585.

English, S.A., Wilkinson, C.R., and Baker, V.J. (Eds.). *Survey Manual for Tropical Marine Resources*, 2013, p. 390.

Fegl, F., Marins, J.E.R., and Costa Neto, Cl. Unpublished studies; see also Frehden, O. and Goldschmidt, L. *Microchim. Acta*, 2 (1937): 184.

Feigle, F. and Anger, V. In: *Spot Test in Organic Analysis*, 7th edn. Translated by R.E. Oesper. Completely Revised and Enlarged. Elsevier Scientific, Amsterdam, the Netherlands, 1966.

Figueroa, L.A. and Silverstein, J.A. Ruthenium red adsorption method for the measurement of extracellular polysaccharides in sludge flocs. *Biotechnol. Bioeng.*, 33 (1989): 941–947.

Filisetti-Cozzi, T.M.C.C. and Carpita, N.C. Measurement of uronic acid without interference from neutral sugars. *Anal. Biochem.*, 197(1) (1991): 157–162.

Foulger, J.H., *J. Biol. Chem.*, 99 (1932): 207; Foulger, J.H., *Compt. Rend.*, 196 (1933): 2984.

Frengova, G.I., Simova, E.D., Beshkova, D.M., and Simov, Z.I. Production and monomer composition of exopolysaccharides by yogurt starter cultures. *Can. J. Microbiol.*, 46 (2000): 1123–1127.

Gaecia-Garibay, M. and Marshall, V.M.E. Polymer production by *Lactobacillus delbulgaricus*. *J. Appl. Bacteriol.*, 70 (1991): 325–328.

Gehrke, T., Telegdi, J., Thierry, D., and Sand, W. Importance of extracellular polymeric substances from *Thiobacillus ferrooxidans* for bioleaching. *Appl. Environ. Microbial.*, 64 (1998): 2743–2747.

Ghebregzabher, M.S., Ruffini, S., Bonaldi, B., and Lato, M. Thin-layer chromatography of carbohydrates. *J. Chromatogr. A*, 127 (1976): 133–162.

Gilmour, J., Smith, L., Cook, K., and Pincock, S. *Discovering Scott Reef: 20 Years of Exploration and Research*, 2013, pp. 179.

Gol Mohammad, F., Eikani, M.H., and Shokr Elahzadeh, S. Review on extraction of medicinal plants constituents by superheated water. *J. Med. Plants*, 7 (2008): 1–24.

Gremm, J. and Kaplan, L.A. Dissolved carbohydrates in steamwater determined by HPLC and pulsed amperometric detection. *Limnol. Oceanogr.*, 42 (1997): 384–393.

Grobben, G.J., Sikkema, J., Smith, M.R., and de bont, J.A.M. Production of extracellular polysaccharides by *Lactobacillus delbrueckii* ssp. Bulgaricus NCFB2772 grown in a chemically defined medium. *J. Appl. Bacteriol.*, 79 (1995): 103–107.

Guilbaut, G.G. *Analytical Uses of Immobilized Enzymes.* Marcel Dekker, New York, 1984.

Handa, N. and Tominaga, H. A detailed analysis of carbohydrates in marine particulate matter. *Mar. Biol.*, 2 (1969): 228–235.

Hanson, R.B. and Snyder, J. Enzymatic determination of glucose in marine environments: Improvement and note of caution. *Mar. Chem.*, (1979): 353–362.

Harding, I.P., Marshall, V.M., Elvin, M., Gu, Y., and Laws, A.P. Structural characterization of a perdeuteriomethylated exopolysaccharide by NMR spectroscopy: Characterization of the novel exopolysaccharide produced by *Lactobacillus delbrueckii* subsp. Bulgaricus EU23. *Carbohydr. Res.*, 338 (2003): 61–67.

Hernes, P.J., Hedges, J.I., Peterson, M.L., Wakeham, S.G., and Lee, C. Neutral carbohydrates geochemistry of particulate material in the central equatorial Pacific. *Deep-Sea Res. II.*, 43 (1996): 1181–1204.

Herrrero, M., Martin-Alvarez, P.J., Senorans, F.J., Cifuentes, A., and Ibanez, E. Optimization of accelerated solvent extraction on antioxidants from *Spirulina platensis* microalga. *Food Chem.*, 93 (2005): 417–423.

Herrero, M.A., Cifuentes, A., and Ebanez, E. Sub and supercritical fluid extraction of functional ingredients from different natural sources: Plants, food-by-products, algae and microalgae: A review. *Food Chem.*, 98 (2006): 136–148.

Hicks, R.A., Owen, C.J., and Aas, P. Deposition, resuspension, and decomposition of particulate organic matter in sediments of Lake Itasca, Minnesota, USA. *Hydrobiologia*, 284 (1994): 79–91.

Hicks, S.E. and Carey, R.G. Glucose determination of carbohydrates in seawater. *Limnol. Oceanogr.*, 13 (1968): 371–363.

Honda, S. Post column derivatization for chromatographic analysis of carbohydrates. *J. Chromatogr. A*, 720 (1996): 183–199.

Hung, C.C. and Santschi, P.H. Spectrophotometric determination of total uronic acids in seawater using cation exchange separation and preconcentration by lyophilization. *Anal. Chim. Acta*, 427 (2001): 111–117.

Hung, C.C., Santschi, P.H., and Gillow, J.B. Isolation and characterization of extracellular polysaccharides produced by *Pseudomonas fluorescence* Biovar II. *Carbohydr. Polym.*, 61 (2005): 141–147.

Hung, C.-C., Tang, D., Warnken, K.W., and Santschi, P.H. Distributions of carbohydrates, including uronic acids, in estuarine waters of Galveston Bay. *Mar. Chem.*, 73 (2001): 305–318.

Hung, C.C., Warken, K.W., and Santschi, P.H. A seasonal survey of carbohydrates and uronic acids in the Trinity River, Texas. *Org. Geochem.*, 36 (2005): 463–474.

Hwang, H.J., Kim, S.W., Xu, C.P., Choi, J.W., and Yun, J.W. Production and molecular characteristics of four groups of exopolysaccharides from submerged culture of *Phellinus gilvus*. *J. Appl. Microbiol.*, 94 (2003): 708–719.

Ibanez, E., Herrero, M., Mendiola, J.A., and Castro-Puyana, M. Extraction and characterization of bioactive compounds with health benefits from marine resources: Macro and micro algae, cyanobacteria, and invertebrates. In: *Marine Bioactive Compounds: Sources, Characterization and Applications*. Hayes, M., Ed. Springer Science + Business Media, New York, 2012, doi 10.1007/978-1-4614-1247-2_2.

Ittekkot, M.V., Degens, E.T., and Brockmann, U. Monosaccharide composition and acid-hydrolyzable carbohydrates in particulate organic matter during a plankton bloom. *Limnol. Oceanogr.*, 27 (1982): 770–776.

Ittekkot, V., Degens, E.T., and Honjo, S. Seasonality in the fluxes of sugars, amino acids and amino sugars to the deep ocean: Panama Basin. *Deep-Sea Res.*, 31 (1984b): 1071–1083.

Ittekkot, V., Deuser, W.G., and Degens, E.T. Seasonality in the fluxes and sugars, amino acids and amino sugars to the deep ocean: Sargasso Sea. *Deep Sea Res.*, 31 (1984a): 1057–1069.

Isaac, G.M., Waldeback, M., Eriksson, U., Odham, G., and Markides, K.E. Total lipid extraction of homogenized and intact lean fish muscles using pressurized fluid extraction techniques. *J. Agric. Food Chem.*, 53 (2005): 5506–5512.

Johnson, D.C. and LaCourse, W.R. Liquid chromatography with pulse amperometric detection at gold and platinum electrodes. *Anal. Chem.*, 62 (1990): 589–597.

Johnson, K.M., Burney, C.M., and Sieburth, J.McN. Doubling the production and precision of the MBTH Spectrophotometric assay for dissolved carbohydrates in seawater. *Mar. Chem.*, 10 (1981): 467–473.

Johnson, K.M. and Sieburth, J.McN. Dissolved carbohydrates in seawater. 1. A precise spectrophotometric method for monosaccharides. *Mar. Chem.*, 5 (1977): 1–13.

Jorgensen, N.O.G. and Jensen, R.E. Microbial fluxes of free monosaccharides and total carbohydrates in freshwater determined by PAD-HPLC. *FEMS Microbiol. Ecol.*, 14 (1994): 79–94.

Josefesson, B. Determination of soluble carbohydrates in seawater by partition chromatography after desalting by ion-exchange membrane electrodialysis. *Anal. Chim. Acta*, 52 (1970): 65–73.

Kailemia, M.J., Ruhaak, L.R., Lebrilla, C.B., and Amster, I.J. Oligosaccharide analysis by mass spectrometry: A review of recent development. *Anal. Chem.*, 86 (2014): 196–212.

Kang, K., Quitain, A.T., Diamon, H. et al. Optimization of amino acids production from waste fish entrails by hydrolysis in sub- and supercritical water. *Can. J. Chem. Eng.*, 79 (2001): 65–70.

Kerherve, P., Buscail, B., Gadel, F., and Serve, L. Neutral monosaccharides in surface sediments of the northwestern Mediterranean Sea. *Org. Geochem.*, 33 (2002): 421–435.

Kerherve, P., Charriere, B., and Gabel, F. Determination of marine monosaccharides by high-pH anion exchange chromatography with pulsed amperometric detection. *J. Chromatogr. A*, 718 (1995): 283–289.

Kerherve, P., Charrierre, B., Stavrakakis, S., Ferrand, J.L., Monaco, A., and Delsaut, N. Biogeochemistry and dynamics of setting particles fluxes at the Antikythira Strait (Eastern Mediterranean). *Prog. Oceanogr.*, 44 (1999): 651–675.

Kirchman, D.L., Meon, B., Ducklow, H.W., Carlson, C.A., Hansell, D.A., and Steward, G. Glucose fluxes and concentrations of dissolved combined sugars (polysaccharides) in the Ross Sea and polar front zone, Antarctica. *Deep-Sea Res.*, II 48 (2001): 4179–4197.

Klok, J.H., Cox, H.C., Bass, M., Schuyal, P.J.W., de leeuw, J.W., and Schenck, P.A. Carbohydrates in recent marine sediments-I. Origin and significance deoxy- and O-methyl-monosaccharides. *Org. Geochem.*, (1984): 73–84.

Knoshaug, E.P., Ahlgren, J.A., and Trempy, J.E. Growth associated exopolysaccharide expression in *Lactococcus lactis* subspecies cremoris Ropy 352. *J. Dairy Sci.*, 83 (2000): 633–640.

Kuhn, R. and Jerchel, D. *Chem. Ber.*, 74 (1941): 949.

Larsson, L.T. and Samuelson, O. An automated micromethod for the separation of monosaccharides by the partition chromatography. *Microchim. Acta*, 2 (1967): 328–332.

Lee, Y.C. High-performance anion-exchange chromatography for carbohydrates analysis. *Anal. Biochem.*, 189 (1990): 151–162.

Leppard, G.G. The characterization of algal and microbial mucilage and their aggregates in aquatic ecosystem. *Sci. Total Environ.*, 165 (1995): 103–131.

Leppard, G.G. Colloidal organic fibrils of acid polysaccharides in surface water: Electron optical characteristics, activities and chemical estimates of abundance. *Colloids Surf. A Physiochem. Eng. Asp.*, 1230 (1997): 1–15.

Levander, F., Svensson, M., and Radstrom, P. Small-scale analysis of exopolysaccharides from *Streptococcus thermophilus* grown in a semi-defined medium. *BMC Microbiol.*, September 1–23, 2001.

Li, B.W. and Andrews, K.W. Separation of trimethylsilylated oximes of monosaccharides by capillary gas-liquid chromatography. *Chromatographia*, 21 (1986): 596–598.

MacCarthy, M., Hedges, J.I., and Benner, R. Major biochemical composition of dissolved high-molecular-weight organic matter in seawater. *Mar. Chem.*, 55 (1996): 281–297.

Marshall, V.M., Cowie, E.N., and Moreton, R.S. Analysis and production of two exopolysaccharides from *Lactococcus lactis* subsp. Cremoris LC30. *J. Dairy Res.*, 62 (1995): 621–628.

Mendiola, J.M., Herroero, M., Cifuentes, A., and Ibanez, E. Use of compressed fluids for sample preparation: Food applications. *J. Chromatogr. A*, 1152 (2007): 234–246.

Meyer, A., Raba, C., and Fischer, K. Ion-pair RP-HPLC determination of sugars, amino sugars, and uronic acids after derivatization with p-amino benzoic acid. *Anal. Chem.*, 73 (2001): 2377–2382.

Momose,T. and Ohkura,Y., Organic analysis. XX. Microestimation of blood sugar with 5-hydroxy-1-tetralone. *Chem. Pharm. Bull.*, Tokyo, Japan, 4 (1956): 209.

Momose, T., Okhura, Y. Organic analysis. XIII. Estimation of hexose with 5-hydroxy-l-tetralone. *Chem. Pharm. Bull.*, Tokyo, Japan, 7 (1959): 31.

Momose, T., Okhura, Y. Organic analysis. X. Reaction mechanism of 5-hydroxytetralone with glucose. *Chem. Pharm. Bull.*, Tokyo, Japan, 6 (1958): 412.

Mopper, K. Sugars and uronic acids in sediment and water from the Black Sea and North Sea emphasis on analytical techniques. *Mar. Chem.*, 5 (1977): 585–603.

Mopper, K. Improved chromatographic separations on anion-exchange resins. I. Partition chromatography of sugars in ethanol. *Anal. Biochem.*, 85 (1978a): 528–532.

Mopper, K. Improved chromatographic separations on anion-exchange resins. II. Separation of uronic acids in acetate medium and detection with a noncorrosive reagent. *Anal. Biochem.*, 86 (1978b): 597–601.

Mopper, K. Improved chromatographic separation on anion-exchange resins. III. Sugars in borate medium. *Anal. Biochem.*, 87 (1978c): 162–168.

Mopper, K. and Degens, E.R. New chromatographic sugar autoanalyzer with a sensitivity 10–10 mol. *Anal. Biochem.*, 45 (1972): 147–153.

Mopper, K. and Gindler, E.M. A new noncorrosive dye reagent for automated sugar chromatography. *Anal. Biochem.*, 56 (1973): 440–442.

Mopper, K. and Johnson, L. Reversed-phase liquid chromatography analysis of DNS-sugars. Optimization of derivatization and chromatographic procedures and applications to natural samples. *J. Chromatogr.*, 256 (1983): 27–38.

Mopper, K., Dawson, G., Liebezelt, G., and Hansen, H.P. Borate complex ion exchange chromatography with fluorimetric detection for determination of saccharides. *Anal. Chem.*, 52 (1980): 2018–2022.

Mopper, K., Schultz, C.A., Chvolot, L., Germain, C., Revuelta, R., and Dawson, R. Determination of sugars in unconcentrated seawater and other natural waters by liquid chromatography. *Environ. Sci. Technol.*, 26 (1992): 133–137.

Morris, D.L. Determination of carbohydrates with Dreywood's anthrone reagent. *Science*, 107 (1948): 254.

Mozzi, F., Giori, G.S., Oliver, G., and de Valdez, G.F. Exopolysaccharide production by *Lactobacillus casei* in milk under different growth conditions. *Milchwissenschaft*, 51 (1996): 670–673.

Nairn, A.B., York, W.S., Harris, K., Hall, E.M., Pierce, J.M., and Moremen, K.W. Regulation of glycans structures in murine embryonic stem cells: Combined transcript of profiling of glycans-related gens and glycans structural analysis. *J. Biol. Chem.*, 283 (2008): 1798–17313.

Nieto, A., Borrull, F., Pocurull, E., and Marce, R.M. Pressurized liquid extraction: A useful technique to extract pharmaceuticals and personal-care products from sewage sludge. *Trends Anal. Chem.*, 29 (2010): 752–764.

Ogier, S., Disnar, J.R., Alberic, P., and Bourdier, G. Neutral carbohydrate geochemistry of particulate matter (trap and core sediment) in a eutrophic lake (Aydat France). *Org. Geochem.*, 32 (2001): 38–50.

Ohtsubo, K. and Marth, J.D. Glycosylation in cellular mechanisms of health and disease. *Cell*, 126 (2006): 855–867.

Olechno, J., Carter, S.R., Edwards, W.T., and Gillen, D.G. Developments in the chromatographic determination of carbohydrates. *Am. Biotechnol. Lab.*, 5 (1987): 38–50.

Ong, E.S., Cheong, J.S.H., and Goh, D. Pressurized hot water extraction bioactive or marker compounds in botanicals and medicinal plant materials. *J. Chromatogr. A*, 1112 (2006): 92–102.

Panagiotopoulos, C. and Sempere, R. Analytical methods for the determination of sugars in marine samples: A historical perspective and future directions. *Limnol. Oceanogr. Meth.*, 3 (2005a): 419–454.

Panagiotopoulos, C. and Sempere, R. The molecular distribution of combined aldoses in sinking particles in various oceanic conditions. *Mar. Chem.*, 95 (2005b): 31–49.

Passow, U. Transparent exopolymer particles (TEP) in aquatic environments. *Prog. Oceanogr.*, 55 (2002): 287–333.

Passow, U. and Alldredge, A.L. A due binding assay for the Spectrophotometric measurement of transparent exopolymer particles (TEP). *Limnol. Oceanogr.*, 40 (1995): 1326–1335.

Petry, S., Furlan, S., Crepeau, M.J., Cerning, J., and Desmazeaud, M. Factors affecting exocellular polysaccharide production by *Lactobacillus delbrueckii* subsp. bulgaricus grown in a chemically defined medium. *Appl. Environ. Microbiol.*, 66 (2000): 3427–3431.

Plunket, M.A. The quantitative determination of some organic compounds in marine sediments. *Deep-Sea Res.*, 4 (1957): 259–262.

Prescher, J.A. and Bertozzi, C.R. Chemical technologies for probing glycans. *Cell*, 126 (2006): 851–854.

Prouty, A.M. and Gunn J.S. Comparative analysis of *Salmonella enterica* serovar typhimurium biofilms formation on gallstones and on glass. *Infect. Immun.*, 71 (2003): 7154–7158.

Quigley, M.S., Santschi, P.H., Guo, L., and Honeyman, B.D. Sorption reversibility and coagulation behavior of 234Th with surface-active marine organic method. *Mar. Chem.*, 76 (2001): 27–45.

Quigley, M.S., Santschi, P.H., Hung, C.C., Guo, L., and Honeyman, B.D. Importance of acid polysaccharides for Th(234) complexation to marine organic matter. *Limnol. Oceanogr.*, 47(2) (2002): 367–377.

Rathore, H.S. Methods of and problems in analyzing pesticide residues in the environment. In: *Handbook of Pesticides: Methods of Pesticide Residues Analysis*. Leo, M., Nollet, L., and Rathore, H.S., Eds. CRC Press, Taylor &Francis Group, Boca Raton, FL, 2010, pp. 7–46.

Rathore, H.S. and Khan, A.A. Fungicide and herbicide residues in water. In: *Handbook of Water Analysis*. Nollet, L.M.L., Ed. Marcel Dekker, New York, 2000, pp. 609–654.

Ray, B. Polysaccharides from *Enteromorpha compressa*: Isolation, purification and structural features. *Carbohyd. Polym.*, 66 (2006): 408–416.

Ricciardi, A., Parente, E., Crudele, M.A., Zanetti, F., Scolari, G., and Minnazzu, I. Exopolysaccharide production by *Streptococcus thermophilus* SY: Production and preliminary characterization of the polymer. *J. Appl. Micribiol.*, 92 (2002): 297–306.

Rich, J.H., Ducklow, H.W., and Kirchman, D.L. Concentrations and uptakes of neutral monosaccharides along 140°W in the equatorial Pacific: Contribution of glucose to heterotrophic bacterial activity and the DOM flux. *Limnol. Oceanogr.*, 41 (1996): 595–604.

Rich, J.H., Gosselin, M., Sherr, E., Sherr, B., and Kirchman, D.L. High bacterial production, uptake and concentrations of dissolved organic matter in the Central Arctic Ocean. *Deep-Sea Res.*, II 44 (1997): 1645–1663.

Richter, B.E., Jones, B.A., Ezell, J.L., Porter, N.L., Audalovic, N., and Pohl, C. Accelerated solvent extraction: A technique for sample preparation. *Anal. Chem.*, 68 (1996): 1033–1039.

Rimada, P.S. and Abraham, A.G. 2001. Polysaccharide production by kefir grains during whey fermentation. *J. Dairy Res.*, 68 (2001): 653–661.

Rocklin, R.D. and Pohl, C.A. Determination of carbohydrates by anion-exchange chromatography with pulsed amperometric detection. *J. Liq. Chromatogr.*, 6 (1983): 1577–1590.

Ruas-Madeodo, P., Tuinier, R., Kanning, M., and Zoon, P. Role of exopolysaccharide produced by *Lactococcus lactis* subsp. cremoris on the viscosity of fermented milks. *Int. Dairy J.*, 12 (2002): 689–695.

Ruas-Madiedo, P. and de los, R.C.G. Invited review: Methods for the screening isolation, and characterization of exopolysaccharide produced by lactic acid bacteria. *J. Dairy Sci.*, 88 (2005): 843–856.

Sakugawa, H. and Handa, N. Chemical studies on dissolved carbohydrates in water samples collected from North Pacific and Bering Sea. *Oceanol. Acta*, 8 (1985): 185–196.

Santschi, P.H., Balnois, E., Wilkinson, K.J., Zhang, J., Buffle, J., and Guo, L. Fibrillar polysaccharides in marine macromolecular organic matter as imaged by atomic force microscopy and transmission electron microscopy. *Limnol. Oceanogr.*, 43(5) (1998): 896–908.

Santschi, P.H., Hung, C.C., Guo, L., Pinckney, J., Schultz, G., Alvarado-Quiroz, N., and Walsh, I. Control of acid polysaccharide production, and Th(234) and POC. Export fluxes by marine organisms. *Geophys. Res. Lett.*, 30(2) (2003): Art. No. 1044, doi 10.1029/2002GL016046.

Senior, W., Chevelot, L., and Courtot, P. Determination of carbohydrates in the marine environment by HPLC. *J. Rech. Oceanogr.*, 10 (1985): 105–107.

Skoog, A. and Benner, R. Aldoses in various size fractions of marine organic matter: Implications for carbon cycling. *Limnol. Oceanogr.*, 42 (1997): 1803–1813.

Spiric, A.D., Trbovic, D., Vrabic, D., Djinovic, J., Petronijevic, R., and Matekalo-Sverak, V. Statistical evaluation of fatty acid profile and cholesterol content in fish (common carp) lipids obtained by different samples preparation procedures. *Anal. Chim. Acta*, 672 (2010): 66–71.

Teo, C.C., Tan, S.N., Yong, J.W.H., Hew, C.S., and Ong, E.S. Pressurized hot water extraction (PHWE). *J. Chromatogr. A*, 1217 (2010): 2484–2494.

Torino, M.I., Mozzi, F., Sesma, F., and de Valdez, G.F. Effect of stirring on growth and phosphopolysaccharide production by *Lactobacillus helveticus* ATCC 15807 in milk. *Milchwissenschaft*, 55 (2000a): 204–206.

Torino, M.I., Sesma, F., and de Valdez, G.F. Semi-defined media for the exopolysaccharide (EPS) production by *Lactobacillus helveticus* ATCC 15807 and evaluation of the components interfering with the EPS quantification. *Milchwissenschaft*, 55 (2000b): 314–316.

Torino, M.I., Taranto, M.P., Sesma, F., and de Valdez, G.F. Heterofermentative pattern and exopolysaccharide production by *Lactobacillus helveticus* ATCC 15807 in response to environmental pH. *J. Appl. Microbiol.*, 91 (2001): 846–852.

Turner, C. and Ibanez, E. Pressurized hot water extraction. In: *Enhancing Extraction Processes in the Food Industry*. Levoka, N., Vorobiev, E., and Chemat, F., Eds. Taylor & Francis Group, LLC, Boca Raton, FL, 2011.

Van Geel-Schutten, G.H., Gaber, E.J., Smit, E., Bonting, K., Smith, M.R., ten Brink, B., Kamerling, J.P., Vliegenthart, J.F.G., and Dijkhuizen, L. Biochemical and structural characterization of the glucan and fructan exopolysaccharides synthesized by *Lactobacillus reuteri* wild-type strain and by mutant strains. *Appl. Environ. Microbiol.*, 65 (1999): 3008–3014.

van Kranenburg, R., Marugg, J.D., van Swam, I.L., Willem, N.J. and de Vos, W.M. Molecular characterization of the plasmid-encoded eps gene cluster essential for exopolysaccharide biosynthesis in *Lactococcus lactis*. *Mol. Microbiol.*, 24 (1997): 387–397.

Van Marle, M.E. and Zoom, P. Permeability and rheological properties of microbially and chemically acidified skim-milk gels. *Neth. Milk Dairy J.*, 49 (1995): 47–65.

Veningelgem, F., Zamfir, M., Mozzi, F., Adriany, T., Vancanney, M., Swing, J., and De vuyst, L. Biodiversity of exopolysaccharides produced by *Streptococcus thermophilus* strains is reflected in their production and their molecular and functional characteristics. *Appl. Environ. Microbiol.*, 70 (2004): 900–912.

Walter, J.S. and Hedges, J.I. Simultaneous determination of uronic acids and aldoses in plankton, plant tissues and sediments capillary gas chromatography of N-hexylaldonamide and alditols acetate. *Anal. Chem.*, 60 (1988): 988–994; *Acta*, 63: 4132–4425.

Wang, L.J. and Weller, C.L. Recent advances in extraction of nutraceuticals from plants. *Trends Food Sci. Technol.*, 17 (2006): 300312.

Wingender, J., Neu, R.R., and Flemming, H.C. *Microbial Extracellular Polymeric Substances: Characterization, Structures and Function.* Springer, Berlin, Germany, 1999.

Witter, A.E. and Luther III, G.W. Spectrophotometric measurement of seawater carbohydrates concentrations in neritic and oceanic waters from the U.S. Middle Atlantic Bight and the Delaware Estuary. *Mar. Chem.*, 77 (2002): 143–156.

Wu, Y., Pan, Y., and Sun, C. Isolation, purification and structural investigation of a water-soluble polysaccharide from *Solanum lyratum* Thunb. *Int. J. Biol. Macromol.*, 36 (2005): 241–145.

Xu, C. Optimized procedures for extraction, purification and characterization of exopolymeric substances (EPS) from two bacteria (*Sagittula stellata* and *Pseudomonas fluorescens* Biovar II) with relevance to the study of actinide binding in aquatic environment. A Thesis. Submitted to the Office of Graduate Students of Texas A&M University. In partial fulfillment of the required for the degree of Master of Science on Major Subject: Oceanography, Galveston, TX, 2007.

Yang, X.B., Gao, X.D., Han, F., Xu, B.S., Song, Y.C., and Tan, R.X. Purification, characterization and enzymatic degradation of YCP, a polysaccharide from marine filamentous fungus *Phoma herbarum* YS4108. *Biochimie*, 87 (2005): 741–754.

Yang, Z., Huttunen, E., Staff, M., Widmalm, G., and Tenhu, H. Separation, purification and characterization of extracellular polysaccharides produced by slime-forming *Lactococcus lactis* ssp. cremoris strains. *Int. Dairy J.*, 9 (1999): 631–638.

Ying, Z., Han, X., and Li, J. Ultrasound-assisted extraction of polysaccharides from mulberry leaves. *Food Chem.*, 127 (2011): 1273–1279.

Yoshida, H., Takahashi, Y., and Terashima, M. A simplified reaction model for production of oil, amino acids and organic acids from fish meat by hydrolysis under sub-critical and super-critical conditions. *J. Chem. Eng. Jpn.*, 36 (2003): 441–448.

Yoshida, H., Terashima, M., and Takahashi, Y. Production of organic acids and amino acids from fish meat by super critical water hydrolysis. *Biotechnol. Prog.*, 15 (1999): 1090–1094.

Zappert, R. Unpublished studies, 2016.

CHAPTER 4

Understanding the Role of Cell Disruption Methods in Extracting Lipids

Avinesh R. Byreddy and Munish Puri

CONTENTS

4.1 Introduction   61
4.2 Selection of Suitable Extraction Solvents   63
4.3 Extraction of Lipids from Dry Biomass   63
4.4 Lipid Extraction from Wet Biomass   66
4.5 Microalgae Cell Disruption for Lipid Extraction   66
4.6 Effect of Cell Disruption Methods on Lipid Extraction   67
4.7 Perspectives   70
References   70

4.1 INTRODUCTION

Biodiesel is emerging as an alternative fuel for diesel engines, which refers to long-chain alkyl fatty acid esters. The increase in petroleum prices and the environmental benefits have resulted in an increased production of biodiesel. Biodiesel occupies 10% of total biofuel production, and the production is about 6 billion L/year globally (Nogueira, 2011). The basic requirements for biodiesel production are feedstock (oil), an alcohol, and a catalyst (e.g., base, acid, or an enzyme). The reaction of biodiesel production occurs in the following steps: production of free fatty acids from triacylglycerols (TAGs) and transesterification of free fatty acids to methanol. This results in the formation of new chemical compounds called methyl esters. The important process variables during the production of biodiesel are reaction temperature, ratio of alcohol to vegetable oil, amount of catalyst, mixing intensity (RPM), raw oils used, and catalyst (Marchetti et al., 2007; Avhad and Marchetti, 2015). In the alkali-based process, generally sodium hydroxide (NaOH) or potassium hydroxide (KOH) is used as the catalyst. During the initial phase of the reaction, the catalyst reacts with the alcohol and forms alkoxy, and this reacts with the TAG to produce biodiesel and glycerol. Glycerol and biodiesel can be separated easily based on gravity. There may be a chance of soap formation by free acid or water contamination that makes the separation process difficult (Gerpen, 2005; Meher et al., 2006). In the acid-based method, an acid with triacylglycerols and alcohol are used. Sulfuric acid is a preferable catalyst in this method (Tran et al., 2013). The acid catalyst gives a high yield of esters, but the reaction rate for conversion requires a long time (Su and Guo, 2014). Although the production cost of biodiesel from the conventional methods such as acid and alkali catalysis is cheap, but associated with some problems, that is, methyl esters yield is normal, recovery of glycerol is difficult and repeated washing is required to purify methyl esters (Gog et al., 2012).

Biodiesel is produced from the renewable energy sources such as biomass. Vegetable oils (edible oils) are also used as a feedstock for the production of biodiesel by transesterification process. The conventional way of biodiesel production is to blend the vegetable oils with the diesel fuel in a suitable ratio, but this direct mixing is technically not possible due to the high viscosity, low stability against oxidation, and low volatility (Robles-Medina et al., 2009). This resulted in an increased price of edible oils and also increase in the production cost of biodiesel that ultimately restricts the usage, although it has more advantages than conventional oils (Hoekman et al., 2012). Sustainable alternative sources should be used to meet the growing demand of oil supply for biodiesel production. In recent times, microalgae have been attractive as feedstock for biodiesel production due to more advantages compared to plants and other microorganisms (Maity et al., 2014; Topf et al., 2014). Although the industrial production of biodiesel from microalgae is still in the developmental stage, scientists are investigating the right policies and strategies to implement alternative feedstocks for industrial biodiesel production.

The major steps in biodiesel production from microalgae are cultivation, harvesting, lipid extraction, and conversion of lipids into fatty acid methyl esters (FAMEs) through transesterification (Figure 4.1). Among these steps, lipid extraction is an energy-intensive and costly process (Mubarak et al., 2015). Generally, the downstream process occupies more than 60% of total biodiesel production cost. The methods used in lipid extraction should be fast, easily scalable, and cost-effective and should not damage the final products. Therefore, it is necessary to reduce the production cost by understanding and developing new lipid extraction methods (Kim et al., 2013). In this write-up, types of solvents used and various methods of extracting lipids from dry and wet biomass are discussed. The overall emphasis on cell disruption methods for extracting lipids is presented.

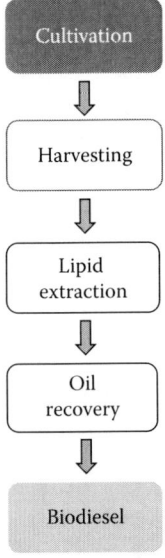

FIGURE 4.1   Downstream processing of lipid extraction from microalgae.

4.2 SELECTION OF SUITABLE EXTRACTION SOLVENTS

In order to optimize lipid extraction yields, solvent selection is an important step. Chloroform/methanol (1:2 v/v) is the most commonly used and reliable organic solvent mixture for total lipid extraction from microalgae (Bligh and Dyer, 1959). Using this organic solvent system, residual endogenous water in the microalgal cells acts as a ternary component that enables the complete extraction of both neutral and polar lipids. This method does not require the complete drying of microalgal biomass. Once the cell debris is removed, more chloroform and water are added to induce biphasic partitioning. The lower organic phase (chloroform) contains most of the lipids (both neutral and polar), while the upper aqueous phase (water with methanol) constitutes most of the nonlipids (proteins and carbohydrates) (Medina et al., 1998). The chloroform and methanol lipid extraction process is fast and quantitative. However, the use of chloroform associated with health, security, and regulatory problems and as such is undesirable. Folch et al. originally developed the chloroform and methanol lipid extraction procedure to extract lipids from brain tissue (Folch et al., 1951). The lipid extraction yield depends on the solvent mixture used in the extraction process. As a result, its efficiency in lipid extraction from microalgal biomass need to be further investigated. The same solvent mixture (chloroform and methanol) is not suitable to extract total lipids from different microalgae samples. Therefore, the extraction solvent system should be selective for lipid extraction and moreover nonlipid compounds also get extracted (Amaro et al., 2011).

Jeon et al. examined 15 different solvents and analyzed their efficiency in lipid extraction from *Chlorella vulgaris* biomass. Lipid extraction yields increased to 25% with methanol/dichloromethane as a solvent mixture compared with other solvents used (Jeon et al., 2013). Mandal et al. also studied the effect of different solvents and their combinations in lipid recovery from *Scenedesmus obliquus*. Around 13% of lipid was extracted using chloroform and methanol (2:1) solvent (Mandal et al., 2013). The efficiency of 13 solvents and their combinations with a range of polarities and solubilities was investigated. The greatest lipid extraction efficiency (11.76%) was obtained upon using a 1:1 mixture of chloroform/ethanol (Ramluckan et al., 2014). Table 4.1 summarizes recent studies investigating the effect of various solvents on lipid extraction.

4.3 EXTRACTION OF LIPIDS FROM DRY BIOMASS

Marine microalgae are single-celled microorganisms with potential industrial applications: as a source of aquaculture feedstock and production of valuable bioactive compounds such as lipids, carotenoids, and enzymes (Amin, 2009; Mercer and Armenta, 2011; Cao et al., 2013). Lipid extraction is the most costly and critical step in biodiesel production. Generally, organic solvent extraction method is most commonly used as it is well established (Folch et al., 1957; Bligh and Dyer, 1959). However, effective extraction of total lipids from microalgae is not possible due to the rigid cell wall. In the case of slow extraction due to the presence of strong cell wall, it is advised to perform an extraction dynamic study in order to improve extraction efficiency (Cho et al., 2012). The most commonly used organic solvents are benzene, hexane, acetone, and chloroform, and these have shown to be effective for lipid extraction from microalgae by disrupting their cell walls (Ramluckan et al., 2014). Selection of the solvent for lipid extraction should take into account its ability to enter into the algal cells efficiently and match the polarity of the desired compounds (for instance, hexane for nonpolar lipids). Physical contact with solvent and lipid material can

TABLE 4.1 Recent Studies Investigating the Use of Organic Solvents in the Extraction of Microalgae Lipids

Marine Microorganisms	Dried Biomass (g)	Organic Solvents (v/v)	Duration of Reaction Conditions (min)/Agitation (rpm)/Temp. (°C)	Wt% of Total Extracted Lipids	Reference
Schizochytrium S31	0.05	Chloroform/methanol (2:1)	NA	22	Byreddy et al. (2015)
Chlorella sp.	1	Chloroform/ethanol (1:1)	3 h/NA/boiling point	11.76	Ramluckan et al. (2014)
Scenedesmus obliquus	1	Chloroform/ethanol (2:1)	NA/NA/room temperature	12.9	Mandal et al. (2013)
Nannochloropsis gaditana, Tetraselmis suecica, and *Desmodesmus communis*	0.05	N,N-dimethylcyclohexylamine 50 mg/mL (biomass/solvent)	24 h/stirring/room temperature	29.2 / 57.9 / 31.9	Samori et al. (2013)
Chlorella vulgaris	50	Dichloromethane/methanol (1:1)	1 h/stirring/37	25	Jeon et al. (2013)
Arthrospira platensis	1	Chloroform/methanol (2:1)	NA	20	Baunillo et al. (2012)
Pavlova sp.	10	Water/methanol/ethyl acetate (10:24:48)	NS	44.7	Cheng et al. (2011)
Chlorococcum sp.	4	Hexane, hexane/isopropanol (3:2)	450/800/25	6.8; hexane/isopropanol	Halim et al. (2011)
Synechocystis PCC 6803	15	Chloroform/methanol/water (1:2:0.8)	24 h/stirring/room temperature	4.2	Sheng et al. (2011)

(Continued)

TABLE 4.1 (*Continued*) Recent Studies Investigating the Use of Organic Solvents in the Extraction of Microalgae Lipids

Marine Microorganisms	Dried Biomass (g)	Organic Solvents (v/v)	Duration of Reaction Conditions (min)/Agitation (rpm)/Temp. (°C)	Wt% of Total Extracted Lipids	Reference
Schizochytrium limacinum	1	Water/chloroform/methanol (4:6:12)	NA	57	Johnson and Wen (2009)
Aurantiochytrium sp. strain T66	28–32	Methanol/chloroform/water (2:2:1)	NS/NS/40	55; methanol/ chloroform/water	Jakobsen et al. (2008)
Phaeodactylum tricornutum	10	Ethanol	1440/500/25	6.3	Fajardo et al. (2007)
S. mangrovei IAo-1	100–200	Chloroform/methanol (2:1)	NS/Ns/38	33.2	Leano et al. (2003)
Botryococcus braunii	0.12	Chloroform/methanol (2:1), hexane/isopropanol (3:2), dichloroethane/methanol (1:1), dichloroethane/ethanol (1:1), acetone/dichloromethane (1:1)	50/high/NS	28.6; chloroform/ methanol (2:1)	Lee et al. (1998)
Isochrysis galbana	5	Chloroform/methanol/water (1:2:0.8) hexane/ethanol (1:2.5), hexane/ethanol (1:0.9), butanol, ethanol, EtOH/water (1:1)	60/constant/25	8.9; chloroform/ methanol/water (1:2:0.8)	Grima et al. (1994)
Chaetoceros muelleri and *Monoraphidium minutum*	60	1-butanol, ethanol, hexane/2-propanol (2/3), water/methanol/ chloroform as a control system	90/high (NS)/close to boiling point of each chemical	Control 100% extraction. 2nd highest extraction efficiency with 1-butanol at 94%	Nagle and Lemke (1990)

be achieved by mechanically disrupting the microalgal cells and adding the solvent later (Li et al., 2014). The effect of cell disruption in lipid extraction from microalgae has been reported and shown to significantly improve lipid extraction yields (Lee et al., 2010; Halim et al., 2013; Grimi et al., 2014; Yap et al., 2014; dos Santos et al., 2015).

4.4 LIPID EXTRACTION FROM WET BIOMASS

Drying microalgae biomass prior to lipid extraction accounts for 90% of the total pro-duction energy cost in dry algal lipid extraction for biodiesel production. Microalgae biomass drying is an energy-intensive process; more than 25% of the energy required for this process can be saved by using wet algal biomass (Du et al., 2015). There are few studies on microalgae lipid extraction using wet biomass. Yao et al. investigated the use of isopropanol as a potential solvent in extracting lipids from *Nannochloropsis* sp. wet biomass and achieved 70% of lipid extraction (Yao et al., 2012). Dejoye et al. developed a new environmentally safer method called simultaneous distillation and extraction process for lipid extraction from wet microalgae *Nannochloropsis oculata* and *Dunaliella salina*. This method used alternative solvents such as d-limonene, α-pinene, and p-cymene. The lipid recovery yields were compared with standard Soxhlet extraction and Bligh and Dyer method and observed similar lipid yields (Dejoye et al., 2013). Ethanol as an organic solvent for lipid extraction from wet microalgae *Picochlorum* sp. has been investigated. This study investigated the effect of extraction temperature and time and the influence of solvent biomass ratios on extraction efficiency and lipid class, with results indicating that the extracted lipid was comparable with the Bligh and Dyer method (Yang et al., 2014). Another recent study confirms the suitability and efficiency of ethanol as an extraction solvent for lipid extraction from some microalgae (Yang et al., 2014a). Lipid extrac-tion procedure from wet microalgae biomass using acid and base hydrolysis was recently developed. This method was capable of extracting 79% of the total transesterified lipids from *Chlorella* and *Scenedesmus* sp. (Sathish and Sims, 2012; Sathish et al., 2015).

Recently, oil extraction from wet biomass and direct biodiesel conversion called direct (in situ) transesterification has been investigated (Park et al., 2015). A novel method for microalgal biodiesel production using *Chlamydomonas* sp. JSC4 with 68.7 wt% water content as the feedstock was developed. This method involves microwave disruption, partial dewatering (via the combination of methanol treatment and low-speed centrifuga-tion), oil extraction, and transesterification without removal of solvent and they achieved 97% biodiesel conversion (Chen et al., 2015). The effect of factors influencing the direct biodiesel conversion from wet *Nannochloropsis salina* biomass was studied. The reac-tion conditions such as temperature, reaction time, and solvent and acid quantity were optimized and 90% of total conversion was obtained (Kim et al., 2015).

4.5 MICROALGAE CELL DISRUPTION FOR LIPID EXTRACTION

A large number of lipid extraction procedures are used for microalgal biomass and their extraction efficiencies have been documented. All microalgae are single-celled organisms having an individual cell wall. Furthermore, some species of microalgae have high lipid levels but they are located inside a tough cell wall. The presence of tough cell walls of micro-algae increases the lipid extraction cost since a cell disruption step is required for effective lipid extraction. Cell disruption is an energy-consuming process but an important step in

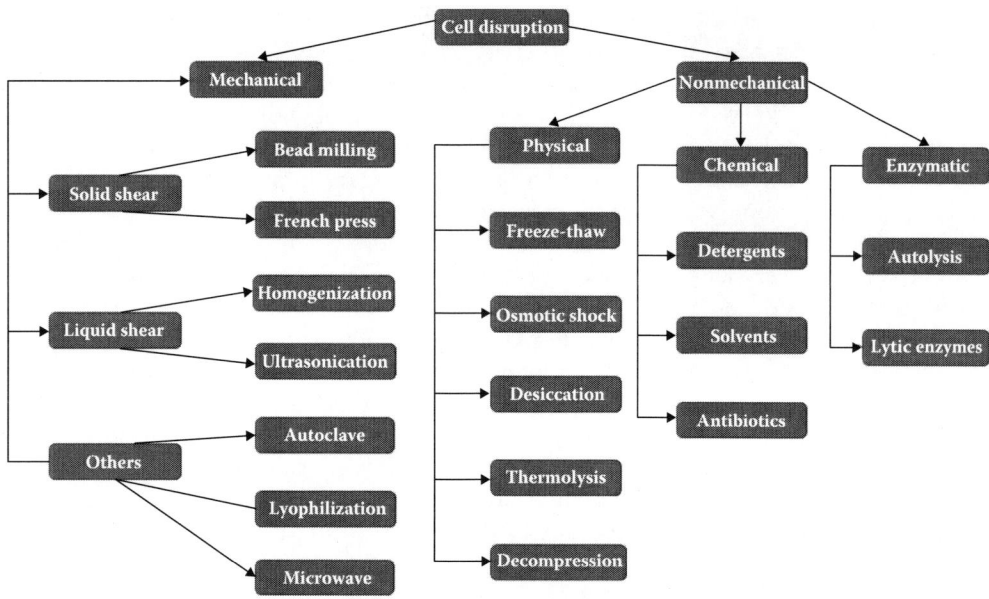

FIGURE 4.2 Classification of cell disruption methods. (Modified from Chisti, Y. and Moo-Young, M., *Enzyme Microb. Technol.*, 8, 194, 1986.)

liberating intracellular lipid molecules from microalgae (Lee et al., 2012). A number of cell disruption methods are available for lipid extraction from microalgae. Different microalgae have different disruption propensities, and as a result, no single method of disruption can be universally applied to the various microalgae species (Halim et al., 2012).

Cell disruption methods (Figure 4.2) that are generally used at the laboratory scale are classified according to the working procedure by which microalgal cells are disrupted, for example, mechanical and nonmechanical (Chisti and Moo-Young, 1986; Middelberg, 1995; Günerken et al., 2015). Mechanical methods include bead mill, press, high-pressure homogenization, ultrasonication, autoclave, lyophilization, and microwave, while nonmechanical methods often involve lysing the microalgal cells with acids, alkalis, enzymes, or osmotic shocks (Günerken et al., 2015).

4.6 EFFECT OF CELL DISRUPTION METHODS ON LIPID EXTRACTION

Microalgal cell wall disruption, in a nonspecific manner, is generally achieved by mechanical methods using solid-shear forces. The basic principle of bead milling involves physically grinding the microalgal cells against the solid surfaces of glass beads in a violent agitation (Chisti and Moo-Young, 1986). Some key factors include high-throughput processing, high biomass loading, good temperature control, easy scale-up procedures, low labor intensity, and high disruption efficiency, which in combination make bead milling a suitable method for implementing large-scale applications (Jahanshahi et al., 2002; Günerken et al., 2015). The method has been used for years to disrupt microorganisms. The size of beads used in the disruption process is important. The optimal diameter of the beads for bacterial cell disruption is 0.1 and 0.5 mm for yeast and other unicellular organisms. The cell disruption efficiency can be further improved by using zirconia-silica-, zirconium oxide–, or titanium carbide–made beads. This is because of their hardness and density. In addition, the separation of beads from the agitated solution is easily achieved due to the high density of beads (Lee et al., 2012).

High-pressure homogenization is an effective and simple cell disruption method. In this process, microalgal suspension pumped through a narrow orifice of a valve under high pressure. Cell disruption is achieved by high-pressure impact (shear forces) of the accelerated fluid jet on the stationary valve surface as well as hydrodynamic cavitation from the pressure drop–induced shear stress (Chisti and Moo-Young, 1986). Some recent studies have observed that high working pressure and cycle number have positive effects on cell disruption efficiency (Halim et al., 2012). In addition, microalgae biomass concentration and species have a major effect on specific energy consumption (Halim et al., 2013). High-speed homogenization is a simple and effective method by which a stirring device can rotate at high rpm within a stator–rotor assembly. The stator–rotor is usually made of stainless steel with a variety of designs. High-speed homogenization achieves cell disruption via hydrodynamic cavitation, generated by stirring at high rpm, and shear forces at the solid–liquid interphase. Hydrodynamic cavitation occurs when the impeller tip reaches approximately 8500 rpm due to the local pressure decreases nearly down to the vapor pressure of the liquid (Kumar and Pandit, 1999; Gogate and Pandit, 2008).

Microalgal cells can be disrupted by an ultrasonication process that is the transmission of sonic waves. An ultrasonicator converts the electrical energy into mechanical vibrations. The probe intensifies the mechanical vibrations, resulting in the formation of pressure waves in the liquids. This mechanism forms the significant amount of cavities (bubbles) in the liquid, which expand during the low-pressure phase and collapse violently during the high-pressure phase. This phenomenon is referred as "cavitation," which releases pressure and heat. Although the cavitational collapse lasts only for a few microseconds, the cumulative effect of bubbles is to release high levels of energy into the liquid (Chisti and Moo-Young, 1986).

Generally, the nonmechanical cell disruption methods involve a use of chemicals, enzymes, and osmotic shock (Monks et al., 2013). These methods primarily depend on the interaction between chemicals or enzymes and cell walls of the microalgae, which modify the outer membrane layer and release the intracellular components. Enzymatic cell disruption is a desirable process due to biological specificity, mild operating conditions, low energy requirements, low capital investment. The use of enzymes normally circumvents the need to use aggressive physical conditions such as high shear stress and elevated temperatures (Choi et al., 2010; Harun and Danquah, 2011). The discovery of the enzyme lysozyme led to interest in the application of lytic enzymes in cell lysis. During enzymatic cell lysis, enzymes bind to cell membranes at specific sites and hydrolyze specific bonds leading to degradation of a cell membrane (Fu et al., 2010). A variety of chemical agents such as antibiotics, chelating agents, chaotropes, detergents, solvents, hypochlorites, acids, and alkali can cause cell disruption. The efficiency of the cell disruption depends on the selectivity of the chemical agent and composition of the microorganisms (Middelberg, 1995; Jalalirad, 2013). A recent study on the comparison of different cell disruption methods on thraustochytrids lipid extraction and their suitability in biodiesel production found a nonmechanical cell disruption by osmotic shocks released maximum lipids (Byreddy et al., 2015).

Lee at al. (2010) assessed the effect of prior mechanical cell disruption on lipid extraction for *Botryococcus* sp., *Chlorella vulgaris*, and *Scenedesmus* sp. using chloroform and methanol (2:1 v/v) as an extraction solvent. Among the methods used (autoclave, bead beating, microwave, sonication, and osmotic shocks), the microwave method resulted in the higher lipid extraction yields in all three microalgal species. Another study by

Zheng et al. (2011) investigated the effect of different mechanical and nonmechanical cell disruption methods for release of lipids from a marine species of *Chlorella vulgaris*. Overall, grinding in liquid nitrogen was identified as the most effective method in terms of disruption efficiency (Zheng et al., 2011).

A similar study was conducted by Halim et al. (2012) in which four cell disruption methods were compared to determine their lipid extraction efficiency. Among the four methods (bead beating, ultrasonication, acid treatment, and high-pressure homogenization), high-pressure homogenization was the most effective in disrupting *Chlorococcum* cells. Summary of recent studies on microalgal cell disruption and their lipid extraction efficiencies are given in Table 4.2.

TABLE 4.2 List of Various Cell Disruption Methods Used for Lipid Extraction

Microalgae	Cell Disruption Method	Maximum Lipid Yield (%)	Reference
Schizochytrium S31	Bead mill	49.4	Byreddy et al. (2016)
Schizochytrium S31	Osmotic shock	48.7	Byreddy et al. (2015)
Thraustochytrium AMCQS5-5		29.1	
Chlorella vulgaris	Ultrasonication	55	dos Santos et al. (2015)
Scenedesmus sp.	Freeze-drying + microwave	29.6	Guldhe et al. (2014)
Chlorella vulgaris	Pressure-assisted ozonation	27	Huang et al. (2014)
Mixed culture	Microwaves	33.7	de Souza Silva et al. (2014)
Chlorella vulgaris	H_2O_2 + $FeSO_4$	17.34	Steriti et al. (2014)
Nannochloropsis sp.	Microwave	38.3	Wahidin et al. (2014)
Chlorella vulgaris	Ultrasonication	52.5	Araujo et al. (2013)
Chlamydomonas reinhardtii	Osmotic shock	NA	Yoo et al. (2012)
Chlorella vulgaris (SAG 211-12)	Grinding + microwaves + sonication	9.82	Šoštarič et al. (2012)
Synechocystis PCC 6803	Microwave + pulsed electric fields	9–13	Sheng et al. (2012)
Chlorella vulgaris	Microwaves	NA	Prommuak et al. (2012)
Scenedesmus sp.	High-pressure homogenization	24.9	Cho et al. (2012)
Synechocystis PCC 6803	Pulsed electric field	25–75	Sheng et al. (2011)
Chlorella sp.	Sonication	21	Prabakaran and Ravindran (2011)
Nostoc sp.	Sonication	18	
Tolypothrix sp.	Microwave	16	
Botryococcus sp.		28.6	Lee et al. (2010)
Chlorella vulgaris	Microwave	10	
Scenedesmus sp.		11	
Scenedesmus dimorphus	Wet milling	25.3	Shen et al. (2009)
Chlorella protothecoides	Bead beater	18.8	

4.7 PERSPECTIVES

Microalgae accumulate a large portion of lipids within their cells compared to terrestrial plants and other microorganisms. Algal biomass is recognized as an alternative feedstock for biodiesel production. Several studies showed that the suitability of microalgae as an alternative renewable source for biofuel production. Hence, selection of a suitable solvent mixture and optimized cell disruption method(s) are important for microalgae lipid extraction and cost-effective biofuel production. Integration of nanotechnology in microalgae lipid extraction and biodiesel production could be one of the possible solutions in economizing processes.

REFERENCES

Amaro, H.M., Guedes, A.C., and Malcata, F.X. Advances and perspectives in using microalgae to produce biodiesel. *Appl. Energy*, 88 (2011): 3402–3410.

Amin, S. Review on biofuel oil and gas production processes from microalgae. *Energy Convers. Manag.*, 50 (2009): 1834–1840.

Araujo, G.S., Matos, L.J.B.L., Fernandes, J.O. et al. Extraction of lipids from microalgae by ultrasound application: Prospection of the optimal extraction method. *Ultrason. Sonochem.*, 20 (2013): 95–98.

Avhad, M.R. and Marchetti, J.M. A review on recent advancement in catalytic materials for biodiesel production. *Renew. Sustain. Energy Rev.*, 50 (2015): 696–718.

Baunillo, K.E., Tan, R.S., Barros, H.R. et al. Investigations on microalgal oil production from *Arthrospira platensis*: Towards more sustainable biodiesel production. *RSC Adv.*, 2 (2012): 11267–11272.

Bligh, E.G. and Dyer, W.J. A rapid method of total lipid extraction and purification. *Can. J. Biochem. Physiol.*, 37 (1959): 911–917.

Byreddy, A.R., Barrow, C.J., and Puri, M. Bead milling for lipid recovery from thraustochytrid cells and selective hydrolysis of *Schizochytrium* DT3 oil using lipase. *Bioresour. Technol.*, 200 (2016): 464–469.

Byreddy, A.R., Gupta, A., Barrow, C.J., and Puri, M. Comparison of cell disruption methods for improving lipid extraction from thraustochytrid strains. *Mar. Drugs*, 13 (2015): 5111–5127.

Cao, H., Zhang, Z., Wu, X., and Miao, X. Direct biodiesel production from wet microalgae biomass of *Chlorella pyrenoidosa* through in situ transesterification. *Biomed. Res. Int.*, 2013 (2013): 6.

Chen, C.L., Huang, C.C., and Ho, K.C. Biodiesel production from wet microalgae feedstock using sequential wet extraction/transesterification and direct transesterification processes. *Bioresour. Technnol.*, 194 (2015): 179–186.

Cheng, C.H., Du, T.B., and Pi, H.C. Comparative study of lipid extraction from microalgae by organic solvent and supercritical CO_2. *Bioresour. Technol.*, 102 (2011): 10151–10153.

Chisti, Y. and Moo-Young, M. Disruption of microbial cells for intracellular products. *Enzyme Microbial Technol.*, 8 (1986): 194–204.

Cho, S.C., Choi, W.Y., and Oh, S.H. Enhancement of lipid extraction from marine microalga, *Scenedesmus* associated with high-pressure homogenization process. *J. Biomed. Biotechnol.*, 2012 (2012): 6.

Choi, S.P., Nguyen, M.T., and Sim, S.J. Enzymatic pretreatment of *Chlamydomonas reinhardtii* biomass for ethanol production. *Bioresour. Technol.*, 101 (2010): 5330–5336.

de Souza Silva, A.P.F, and Costa, M.C. Comparison of pretreatment methods for total lipids extraction from mixed microalgae. *Renew. Energy*, 63 (2014): 762–766.

Dejoye, T.C., Abert, V.M., and Chemat, F. New procedure for extraction of algal lipids from wet biomass: A green clean and scalable process. *Bioresour. Technol.*, 134 (2013): 271–275.

dos Santos, R.R., Moreira, D.M., and Kunigami, C.N. Comparison between several methods of total lipid extraction from *Chlorella vulgaris* biomass. *Ultrason. Sonochem.*, 22 (2015): 95–99.

Du, Y., Schuur, B., Kersten, S.R.A., and Brilman, D.W.F. Opportunities for switchable solvents for lipid extraction from wet algal biomass: An energy evaluation. *Algal Res.*, 11 (2015): 271–283.

Fajardo, A.R., Cerdán, L.E., Medina, A.R. et al. Lipid extraction from the microalga *Phaeodactylum tricornutum*. *Eur. J. Lipid Sci. Technol.*, 109 (2007): 120–126.

Folch, J., Ascoli, I., Lees, M. et al. Preparation of lipid extracts from brain tissue. *J. Biol. Chem.*, 191 (1951): 833–841.

Folch, J., Lees, M., and Stanley, G.H.S. A simple method for the isolation and purification of total lipids from animal tissues. *J. Biol. Chem.*, 226 (1957): 497–509.

Fu, C.C., Hung, T.C., Chen, J.Y. et al. Hydrolysis of microalgae cell walls for production of reducing sugar and lipid extraction. *Bioresour. Technol.*, 101 (2010): 8750–8754.

Gerpen, J.V. Biodiesel processing and production. *Fuel Process. Technol.*, 86 (2005): 1097–1107.

Gog, A., Roman, M., Toşa, M., and Paizs, C. Biodiesel production using enzymatic transesterification—Current state and perspectives. *Renew. Energy*, 39 (2012): 10–16.

Gogate, P.R. and Pandit, A.B. Application of cavitational reactors for cell disruption for recovery of intracellular enzymes. *J. Chem. Technol. Biotechnol.*, 83 (2008): 1083–1093.

Grima, E.M., Medina, A.R., Giménez, A.G. et al. Comparison between extraction of lipids and fatty acids from microalgal biomass. *J. Am. Oil Chem. Soc.*, 71 (1994): 955–959.

Grimi, N., Dubois, A., Marchal, L. et al. Selective extraction from microalgae *Nannochloropsis* sp. using different methods of cell disruption. *Bioresour. Technol.*, 153 (2014): 254–259.

Guldhe, A., Singh, B., Rawat, I. et al. Efficacy of drying and cell disruption techniques on lipid recovery from microalgae for biodiesel production. *Fuel*, 128 (2014): 46–52.

Günerken, E., D'Hondt, E., Eppink, M.H.M. et al. Cell disruption for microalgae biorefineries. *Biotechnol. Adv.*, 33 (2015): 243–260.

Halim, R., Gladman, B., Danquah, M.K., Webley, P.A. Oil extraction from microalgae for biodiesel production. *Bioresour. Technol.*, 102 (2011): 178–185.

Halim, R., Harun, R., Danquah, M.K. et al. Microalgal cell disruption for biofuel development. *Appl. Energy*, 91 (2012): 116–121.

Halim, R., Rupasinghe, T.W.T., Tull, D.L. et al. Mechanical cell disruption for lipid extraction from microalgal biomass. *Bioresour. Technol.*, 140 (2013): 53–63.

Harun, R. and Danquah, M.K. Enzymatic hydrolysis of microalgal biomass for bioethanol production. *Chem. Eng. J.*, 168 (2011): 1079–1084.

Hoekman, S.K., Broch, A., Robbins, C. et al. Review of biodiesel composition, properties, and specifications. *Renew. Sustain. Energy Rev.*, 16 (2012): 143–169.

Huang, Y., Hong, A., Zhang, D. et al. Comparison of cell rupturing by ozonation and ultrasonication for algal lipid extraction from *Chlorella vulgaris*. *Environ. Technol.*, 35 (2014): 931–937.

Jahanshahi, M., Sun, Y., Santos, E. et al. Operational intensification by direct product sequestration from cell disruptates: Application of a pellicular adsorbent in a mechanically integrated disruption-fluidised bed adsorption process. *Biotechnol. Bioeng.*, 80 (2002): 201–212.

Jakobsen, A., Aasen, I., Josefsen, K. et al. Accumulation of docosahexaenoic acid-rich lipid in thraustochytrid *Aurantiochytrium* sp. strain T66: Effects of N and P starvation and O$_2$ limitation. *Appl. Microbiol. Biotechnol.*, 80 (2008): 297–306.

Jalalirad, R. Selective and efficient extraction of recombinant proteins from the periplasm of *Escherichia coli* using low concentrations of chemicals. *J. Ind. Microbiol. Biotechnol.*, 40 (2013): 1117–1129.

Jeon, J.M., Choi, H.W., Yoo, G.C. et al. New mixture composition of organic solvents for efficient extraction of lipids from *Chlorella vulgaris*. *Biomass Bioenergy*, 59 (2013): 279–284.

Johnson, M.B. and Wen, Z. Production of biodiesel fuel from the microalga *Schizochytrium limacinum* by direct transesterification of algal biomass. *Energy Fuel*, 23 (2009): 5179–5183.

Kim, J., Yoo, G., Lee, H. et al. Methods of downstream processing for the production of biodiesel from microalgae. *Biotechnol. Adv.*, 31 (2013): 862–876.

Kim, T.H., Suh, W.I., Yoo, G. et al. Development of direct conversion method for microalgal biodiesel production using wet biomass of *Nannochloropsis salina*. *Bioresour. Technol.*, 191 (2015): 438–444.

Kumar, P.S. and Pandit, A.B. Modeling hydrodynamic cavitation. *Chem. Eng. Technol.*, 22 (1999): 1017–1027.

Leano, E.M., Gapasin, R.S.J., Polohan, B. et al. Growth and fatty acid production of thraustochytrids from Panay mangroves, Philippines. *Fungal Diversity*, 12 (2003): 111–122.

Lee, A.K., Lewis, D.M., and Ashman, P.J. Disruption of microalgal cells for the extraction of lipids for biofuels: Processes and specific energy requirements. *Biomass Bioenergy*, 46 (2012): 89–101.

Lee, J.Y., Yoo, C., Jun, S.Y. et al. Comparison of several methods for effective lipid extraction from microalgae. *Bioresour. Technol.*, 101 (2010) (1, Supplement): S75–S77.

Lee, S., Yoon, B.D., and Oh, H.M. Rapid method for the determination of lipid from the green alga *Botryococcus braunii*. *Biotechnol. Tech.*, 12 (1998): 553–556.

Li, Y., Ghasemi Naghdi, F., Garg, S. et al. A comparative study: The impact of different lipid extraction methods on current microalgal lipid research. *Microbial Cell Factories*, 13 (2014): 14.

Maity, J.P., Bundschuh, J., Chen, C.Y. et al. Microalgae for third generation biofuel production, mitigation of greenhouse gas emissions and wastewater treatment: Present and future perspectives—A mini review. *Energy*, 78 (2014): 104–113.

Mandal, S., Patnaik, R., Singh, A.K. et al. Comparative assessment of various lipid extraction protocols and optimization of transesterification process for microalgal biodiesel production. *Environ. Technol.*, 34 (2013): 2009–2018.

Marchetti, J.M., Miguel, V.U., and Errazu, A.F. Possible methods for biodiesel production. *Renew. Sustain. Energy Rev.*, 11 (2007): 1300–1311.

Medina, A.R., Grima, E.M., Giménez, A.G. et al. Downstream processing of algal poly-unsaturated fatty acids. *Biotechnol. Adv.*, 16 (1998): 517–580.

Meher, L.C., Vidya Sagar, D., and Naik, S.N. Technical aspects of biodiesel production by transesterification—A review. *Renew. Sustain. Energy Rev.*, 10 (2006): 248–268.

Mercer, P. and Armenta, R.E. Developments in oil extraction from microalgae. *Eur. J. Lipid Sci. Technol.*, 113 (2011): 539–547.

Middelberg, A.P.J. Process-scale disruption of microorganisms. *Biotechnol. Adv.*, 13 (1995): 491–551.

Monks, L.M., Rigo, A., Mazutt, M.A. et al. Use of chemical, enzymatic and ultra-sound-assisted methods for cell disruption to obtain carotenoids. *Biocatal. Agric. Biotechnol.*, 2 (2013): 165–169.

Mubarak, M., Shaija, A., and Suchithra, T.V. A review on the extraction of lipid from microalgae for biodiesel production. *Algal Res.*, 7 (2015): 117–123.

Nagle, N. and Lemke, P. Production of methyl ester fuel from microalgae. *Appl. Biochem. Biotechnol.*, 24–25 (1990): 355–361.

Nogueira, L.A.H. Does biodiesel make sense? *Energy*, 36 (2011): 3659–3666.

Park, J.Y., Park, M.S., Lee, Y.C. et al. Advances in direct transesterification of algal oils from wet biomass. *Bioresour. Technol.*, 184 (2015): 267–275.

Prabakaran, P. and Ravindran, A.D. A comparative study on effective cell disruption methods for lipid extraction from microalgae. *Lett. Appl. Microbiol.*, 53 (2011): 150–154.

Prommuak, C., Pavasant, P., Quitain, A.T. et al. Microalgal lipid extraction and evalua-tion of single-step biodiesel production. *Eng. J.*, 16 (2012): 157–166.

Ramluckan, K., Moodley, K.G., and Bux, F. An evaluation of the efficacy of using selected solvents for the extraction of lipids from algal biomass by the soxhlet extraction method. *Fuel*, 116 (2014): 103–108.

Robles-Medina, A., González-Moreno, P.A., Esteban-Cerdán, L. et al. Biocatalysis: Towards ever greener biodiesel production. *Biotechnol. Adv.*, 27 (2009): 398–408.

Samori, C., Lopez Barreiro, D., Vet, R. et al. Effective lipid extraction from algae cultures using switchable solvents. *Green Chem.*, 15 (2013): 353–356.

Sathish, A., Marlar, T., and Sims, R.C. Optimization of a wet microalgal lipid extrac-tion procedure for improved lipid recovery for biofuel and bioproduct production. *Bioresour. Technol.*, 193 (2015): 15–24.

Sathish, A. and Sims, R.C. Biodiesel from mixed culture algae via a wet lipid extraction procedure. *Bioresour. Technol.*, 118 (2012): 643–647.

Shen, Y., Pei, Z., Yuan, W. et al. Effect of nitrogen and extraction method on algae lipid yield. *Int. J. Agric. Biol. Eng.*, 2 (2009): 51–57.

Sheng, J., Vannela, R., and Rittmann, B.E. Evaluation of methods to extract and quantify lipids from *Synechocystis* PCC 6803. *Bioresour. Technol.*, 102 (2011): 1697–1703.

Sheng, J., Vannela, R., and Rittmann, B.E. Disruption of *Synechocystis* PCC 6803 for lipid extraction. *Water Sci. Technol.*, 65 (2012): 567–573.

Šoštarič, M., Klinar, D., Bricelj, M. et al. Growth, lipid extraction and thermal degrada-tion of the microalga *Chlorella vulgaris*. *New Biotechnol.*, 29 (2012): 325–331.

Steriti, A., Rossi, R., Concas, A. et al. A novel cell disruption technique to enhance lipid extraction from microalgae. *Bioresour. Technol.*, 164 (2014): 70–77.

Su, F. and Guo, Y. Advancements in solid acid catalysts for biodiesel production. *Green Chem.*, 16 (2014): 2934–2957.

Topf, M., Koberg, M., Kinel-Tahan, Y. et al. Optimizing algal lipid production and its efficient conversion to biodiesel. *Biofuels*, 5 (2014): 405–413.

Tran, H.L., Ryu, Y.J., and Seong, D. An effective acid catalyst for biodiesel production from impure raw feedstocks. *Biotechnol. Bioprocess Eng.*, 18 (2013): 242–247.

Wahidin, S., Idris, A., Shaleh, S.R.M. et al. Rapid biodiesel production using wet microalgae via microwave irradiation. *Energy Convers. Manage.*, 84 (2014): 227–233.

Yang, F., Long, L., Sun, X. et al. Optimisation of medium using response surface methodology for lipid production by *Scenedesmus* sp. *Mar. Drugs*, 12 (2014a): 1245–1257.

Yang, F., Xiang, W. et al. A novel lipid extraction method from wet *Microalga picochlorum* sp. at room temperature. *Mar. Drugs*, 12 (2014): 1258.

Yao, L., Gerde, J., Wang, T. et al. Oil extraction from microalga *Nannochloropsis* sp. with isopropyl alcohol. *J. Am. Oil Chem. Soc.*, 89 (2012): 2279–2287.

Yap, B.H.J., Crawford, S.A., and Dumsday, G.J. A mechanistic study of algal cell disruption and its effect on lipid recovery by solvent extraction. *Algal Res.*, 5 (2014): 112–120.

Yoo, G., Park, W.K., and Kim, C.W. Direct lipid extraction from wet *Chlamydomonas reinhardtii* biomass using osmotic shock. *Bioresour. Technol.*, 123 (2012): 717–722.

Zheng, H., Yin, J., Gao, Z. et al. Disruption of *Chlorella vulgaris* cells for the release of biodiesel-producing lipids: A comparison of grinding, ultrasonication, bead milling, enzymatic lysis, and microwaves. *Appl. Biochem. Biotechnol.*, 164 (2011): 1215–1224.

CHAPTER 5

Protein Nutraceuticals from Marine Microbes

Lipsy Chopra, Gurdeep Singh, Ramita Taggar,
Raj Kumar, and Debendra K. Sahoo

CONTENTS

5.1	Introduction	76
5.2	Marine Environment: The Widest Front of Bioactive Compounds	76
5.3	Marine Microorganisms as a Source of Bioactive Compounds	77
	5.3.1 Marine Microalgae	78
	5.3.2 Marine Bacteria	79
	5.3.3 Marine Fungi	80
5.4	Marine Microorganisms: Food Grade or Not?	80
5.5	Proteins and Bioactive Peptides from Marine Microbes	81
5.6	Bacteriocins	82
5.7	Peptides: Screening, Isolation, Extraction, Purification, and Analysis	83
	5.7.1 Screening of Cultures and Isolates	83
	5.7.2 Extraction Methods	84
	5.7.2.1 Solvent Extraction	84
	5.7.2.2 Ammonium Sulfate Precipitation	84
	5.7.2.3 pH-Dependent Adsorption–Desorption Method	85
	5.7.2.4 Extraction at Low pH	85
	5.7.2.5 Hydrophobic Resins	86
	5.7.3 Purification of Peptides	86
	5.7.3.1 Ion Exchange Chromatography	86
	5.7.3.2 Gel Permeation	88
	5.7.3.3 Reverse-Phase HPLC	88
	5.7.4 Tricine Polyacrylamide Gel Electrophoresis (Tricine-SDS-PAGE)	90
	5.7.5 Zymography	91
5.8	Quantification and Determination of Composition of Bioactive Peptides	91
5.9	Conclusions	92
References		92

5.1 INTRODUCTION

The term "nutraceuticals," coined from "nutrients" and "pharmaceuticals" by Stephen DeFelici in 1989, is frequently used interchangeably with "functional foods" even though there is a slight disparity between the two. Nutraceuticals are "naturally derived bioactive compounds that are found in foods, dietary supplements and herbal products, and have health promoting, disease preventing, or medicinal properties" (Pandey et al., 2011). Functional foods consist of an ingredient that provides a health-promoting property in addition to its usual nutritional value, for example, probiotic yogurts. When the food is cooked or prepared using "scientific intelligence" with or without understanding of how or why it is being used, the food is called "functional food" (FAO Report, 2007). A functional food that facilitates the prevention or treatment of disorders or diseases other than anemia is called a "nutraceutical." Functional foods supply the required amount of carbohydrates, proteins, vitamins, fats, and so on, needed for healthy survival (FAO Report, 2007). The use of functional foods is becoming an area of growth for the food industry due to the side effects of drugs and negative impact of supplements on human health. Functional foods are one of the largest growing markets in Japan and are defined as regular food derived from naturally occurring ingredients to be consumed as a part of the diet and not in the form of supplements (i.e., in the form of tablets and capsules). However, a thorough investigation of the characteristics and biological activity of functional foods and nutraceuticals, such as their therapeutic or disease-preventing efficacy, proper dosage, and possible adverse effects, is necessary. Significant prospective research in this field includes the following (Jackson and Paliyath, 2011):

1. Identification, quantification, and standardization of promising bioactive components in functional foods
2. Investigation on the effects of functional foods and nutraceuticals on human health
3. Development of strategies to enhance the levels of these compounds in raw and processed foods
4. Establishment of proper dosage and delivery systems
5. Studies on bioavailability and metabolism of functional foods and nutraceuticals
6. Studies on technical and safety issues that have a bearing on Food and Drug Administration (FDA) regulations and health claim evaluations
7. Examination of regulatory issues
8. Research on the stability of the functional foods and nutraceuticals after processing
9. Interaction of functional foods and nutraceuticals with drugs and other functional foods and nutraceuticals

5.2 MARINE ENVIRONMENT: THE WIDEST FRONT OF BIOACTIVE COMPOUNDS

The ocean is the mother of life and it is presumed that the most primeval forms of life originated from this "primordial soup" (Bhatnagar and Kim, 2010). Oceans are regarded as rich in organic compounds that are favorable for the evolution and growth of life. The marine environment covers a broad range of abiotic conditions such as temperature (from below freezing temperature in Antarctic water to about 350°C in deep hydrothermal vents), pressure (1–1000 atm), and nutrients (from oligotrophic to eutrophic), and it has extensive photic and nonphotic zones. This variability has facilitated the extensive speciation at all phylogenetic levels, from microbes to mammals. As these organisms bloom in different types of climatic conditions, they develop certain adaptation mechanisms that may be useful not only for their defense but also for human beings in many ways. The production of bioactive metabolites and compounds is one such adaptation mechanism which helps in the survival from predators. Chemists have been fascinated with the discovery of the first marine natural products due to the enormous structural variability and complexity of metabolites and compounds isolated from marine organisms ranging from plants and invertebrates to marine microorganisms. It was in the 1960s that scientists began to focus on marine environment as an unexplored and novel resource of potentially valuable bioactive metabolites and compounds. This could be due to the fact that more than 95% of the Earth's biosphere is covered with oceans (Davidson, 1995), and researchers are striving to discover bioactive metabolites and compounds in unexpected places as the antibiotic resistance among pathogens increases and the production of novel bioactive compounds has tapered. As a result, more than 10,000 metabolites and compounds have been isolated and characterized from marine environments over the past five decades (Fuesetani, 2000).

The advancement in technology has opened vast areas of research for the extraction of biomedical compounds from the oceans and seas to treat deadly diseases, and it is worth mentioning that marine sources have also exhibited remarkable abilities as producers of bioactives, such as anticancer compounds and secondary metabolites which can act against contagious infections and inflammation. Blunt et al. (2004) have reported that in marine environment, the major sources of biomedical and bioactive compounds are sponges (37%), coelenterates (21%), and microorganisms (18%), followed by algae (9%), echinoderms (6%), tunicates (6%), mollusks (2%), bryozoans (1%), and so on. It is known that about 25% of all pharmaceutical sales are drugs obtained from plant natural products and about 12% are based on natural products derived from microbes (Joffe and Thomas, 1989). However, microorganisms of marine origin have not been given the attention they are worthy of, and an inadequate insight into the capabilities and bioactive potential of these organisms is available in literature to date. There is still scope of research and investigation at a higher magnitude in order to explore the potential of marine organisms including microorganisms as producers of novel metabolites or drugs. In addition to synthetic products, the pharmaceutical industries in most developed and developing nations are now increasingly focusing on the natural products obtained from marine microbes.

5.3 MARINE MICROORGANISMS AS A SOURCE OF BIOACTIVE COMPOUNDS

For the past few decades, soil microbiota has been subject to extensive investigation worldwide; it may be because of the ease of isolation of microorganisms from the lithosphere compared to any other spheres of the Earth (Bhatnagar and Kim, 2010). Due to the huge diversity of novel chemical compounds from these microbes, thousands of strains have been exploited for pharmacological screening. This trend is now altering as the rate of progress in the development of novel drug agents has slowed down, and most of the chemical formulations isolated from terrestrial microbes appear to be repetitive and costly.

Marine microbiota has incredible potential to represent a biomedical resource of unknown magnitude and great promise. It is a well-known fact that marine microorganisms constitute more than 90% of the marine biomass (Delong, 2007) and inhabit all kinds of different environments of the ocean such as deep sea, polar ice, mangroves, hydrothermal vent, and coral reef, and their presence and ecological role vary according to the environmental conditions (Webster and Hill, 2007). As certain groups of microbes have distinctive adaptations for high salt environmental conditions and high hydrostatic pressure conditions, the enormous diversity of the microorganisms in marine habitats is quite noticeable. Owing to the inadequate reports on the culturing techniques and media formulation for culturing marine microbes, pharmaceutical industries have not been able to fully exploit this massive reservoir. Though the myth still prevails that marine microorganisms are difficult to culture, if not uncultivable, a number of reports in the past have proven that marine microorganisms can be cultured successfully (Fenical, 1993; Kobayashi and Ishibashi, 1993), and scientists are exploring unique chemical entities from the marine environment. However, many constraints are associated with marine natural products, particularly the supply of a sufficient quantity of metabolites (Waters et al., 2010; Montaser and Luesch, 2011). There are a number of reports on varied marine microbial community producing metabolites with distinctive structural properties, and these metabolites possess a broad spectrum of pharmaceutical properties such as antimicrobial, anthelmintic, antiplatelet, antituberculosis, antimalarial, antiviral, antiparasitic, antiprotozoal, antitumor, anti-inflammatory, antidiabetic, and anticoagulant effects (Pathirana et al., 1992; Trischman et al., 1994; Imhoff et al., 2011). Major classes of microbes like bacteria and fungi are now the target of biomedical study and fascinating novel metabolites are being produced. In fact, the symbiotic microbial consortia have also proved to be a source of bioactive compounds with pharmaceutical potential. Bacteria and fungi have been sampled over the years from the surfaces of marine plants and the internal tissues of invertebrates and, in particular, marine fungi from these sources appear to be of ever-increasing interest (Cheng et al., 1994; Kakeya et al., 1995). Waters et al. (2010) have emphasized the fact that more than half of the molecules currently in the marine drug development pipeline are highly likely to be produced by microorganisms. In addition, a large number of food-grade metabolites with promising pharmaceutical properties have also been isolated from marine microbes, and there would be an apparent potential to develop those active ingredients as modern nutraceuticals and functional food.

5.3.1 Marine Microalgae

Marine alga has rich biodiversity potential and can provide various requirements as food and bioactive compounds for treatment and remedies for various types of bacterial illness like tuberculosis, viral infections like human immunodeficiency virus and herpes viruses, fungal infections, and protozoan infection like malaria; it is also helpful against helminthic diseases (Chang et al., 2003; Fennell et al., 2003; Goud et al., 2003; Luescher-matti, 2003; Bernan et al., 2004; Maskey et al., 2004; Zhu et al., 2004). Numerous bioactive compounds have been isolated from microalgae. Bioactive sesquiterpenes isolated from the red algae species *Laurencia rigida* has yielded luzonenone, elatol, deschloroelatol, 15-hydroxypalisadin, 3,4-epoxypalisadin, 1,2-dehydro-3,4-epoxypalisadin B, and luzofuran, which have shown antibacterial activity against *Bacillus megaterium* and also possess antifungal activity (Konig et al., 2000; Kuniyoshi et al., 2005). Phlorotannins (phlorofucofuroeckol A, phloroglucinol, eckol, 8.80-beckol, and dieckol) extracted from brown algae, *Ecklonia kurome*, have exhibited antibacterial activity against Gram-positive bacteria such as *Staphylococcus aureus* and *B. cereus* and also against Gram-negative bacteria such as *Campylobacter jejuni*, *E. coli*, *Salmonella enteritidis*, *Salmonella typhimurium*, and *Vibrio parahaemolyticus*. Moreover, in some parts of Japan, phlorotannins are consumed as food so these can be used as food supplements or drugs (Gopal et al., 2008). Bromoditerpenes sphaerolabdadiene-3,14-diol and bromosphaerone isolated from the marine red algae *Sphaerococcus coronopifolius* has exhibited antibacterial action against *S. aureus* and antimalarial activity against the chloroquine-resistant *Plasmodium falciparum* (Etahiri et al., 2001).

5.3.2 Marine Bacteria

The number of reported secondary metabolites from marine bacteria has gradually increased since the last decade (Fenical, 1993; Kobayashi and Ishibashi, 1993; Bernan et al., 1997; Chopra et al., 2014), thus reflecting the increasing attention by groups from academia and industry. The growing requirement for new antimicrobial molecules effective against the resistant strains of microorganisms has encouraged a number of research groups to explore the oceans for novel bioactive compounds. Over the years, massive screening programs have been developed worldwide, and great efforts have been devoted with an aim to isolate novel metabolites from marine bacteria. Marine bacteria are affluent producers of secondary metabolites as they flourish in harsh oceanic climates. The major obstacle in the search for metabolites from marine bacteria is the nonculturability of the majority (over 99%) (Hugenholtz and Pace, 1996). The studies conducted by the scientists at the Scripps Institution of Oceanography showed that marine bacteria are competent of producing unusual bioactive compounds that are not perceived in terrestrial sources (Fenical, 1993). Antagonistic marine bacteria have been isolated from surface and deep waters, but the majority originated from biotic surfaces such as sponges, bryozoans, macroalgae and zooplankton, and corals. Bacteria not only produce metabolites active against other organisms, but they also generate certain compounds that assist in cleaning their environment. Certain marine bacterial species are also rich producers of bioemulsifiers, biosurfactants, and exopolysaccharides. Thermostable proteases, esterases, lipases, and starch- and xylan-degrading enzymes have been actively sought and in many cases are found in bacterial and archaeal hyperthermophilic marine microorganisms (Bertoldo and Antranikian, 2002). Many genus of bacteria have been reported to be the potent producers of bioactive metabolites especially belonging to phylum Actinobacteria, for example, *Streptomyces* spp. Over two-thirds of the clinically useful antibiotics of natural origin (e.g., neomycin, cypemycin, grisemycin, bottromycins, and chloramphenicol) have been produced from them (Buck et al., 1962). Among the several bacteria exhibiting antimicrobial activity, a variant of the ichthyotoxic *Pseudomonas piscicida* showed marked antagonism to various microorganisms. A red-colored bacterium obtained from Puerto Rico was found to excrete vitamin B and antibacterial substances into the seawater (Bhakuni and Rawat, 2005). *Serratia marcescens* has been reported to produce a red-colored antibiotic named prodigiosin (Llagostera et al., 2003; Perez-Tomas et al., 2003). Though prodigiosin exhibited high order of antibiotic and antifungal activities, high toxicity prevented its use as a therapeutic agent. Members of the *Bacillus* group in a broad sense are considered prolific producers of antimicrobial substances, including peptide and lipopeptide antibiotics, and bacteriocins (Stein et al., 2005). The sporulation capability along with the production of antimicrobial compounds confers *Bacillus* strains with an added advantage in terms of their survival in different habitats. Many antibiotics and antimicrobial peptides have been reported from *Bacillus*, for example, sonorensin from *B. sonorensis* (Chopra et al., 2014) and cerein from *B. cereus* (Bizani et al., 2005), and the presence of *Bacillus* species in food does not always entail spoilage or food poisoning, and some species or strains are even used in human and animal food production. For example, *Bacillus subtilis* is used in the production of natto, an East Asian fermented food (Hosoi and Kiuchi, 2003), and a nontoxigenic *B. cereus* subsp. toyoi with probiotic properties is used as an animal feed additive (Lodemann et al., 2008). On the other hand, some *Bacillus* species or strains have been implicated in food poisoning and food spoilage, and these include *B. cereus, Bacillus coagulans, Bacillus mycoides, Bacillus weihenstephanensis, Bacillus licheniformis, B. subtilis, Bacillus pumilus, Bacillus thuringiensis,* and *Bacillus sphaericus* (Jay et al., 2005; Granum, 2007). Therefore, a thorough selection process is essential for the selection and development of *Bacillus* probiotic candidates (Hong et al., 2005) or starter cultures, considering the intraspecies divergent virulence characteristics. The ability to produce antimicrobial peptides is commonly spread among strains of this genus. Because of their capability to produce enzymes, metabolites, and antibiotics, in addition to their physiological properties, *Bacillus* spp. found application in many processes, such as the medical, pharmaceutical, agricultural, and food industries.

5.3.3 Marine Fungi

The world of fungi offers a fascinating and almost endless source of biological diversity, which is a rich source for exploitation (Manoharachary et al., 2005). It has been reported that marine filamentous fungi are sustainable source of bioactive marine natural products (Kobayshi et al., 1996; Namikoshi et al., 2001; Khudyakova et al., 2004). Marine-derived fungal strains primarily produce polyketide-derived alkaloids, peptides, terpenes, and mixed biosynthesis compounds. There are a number of reports on bioactives from marine fungi such as the isolation of gliovictin from marine fungus, *Asteromyces cruciatus* (Shin and Fenical, 1987); cyclosporin A from *Tolypocladium inflatum* in 1976 (approved for clinical application as an immunosuppressant in 1983); and cephalosporin C, a penicillinase-sensitive antibiotic substance active against Gram-negative bacteria from marine fungus *Cephalosporium acremonium* obtained from the sea near a sewage outfall of the coast of Sardinia (Crawford et al., 1952).

5.4 MARINE MICROORGANISMS: FOOD GRADE OR NOT?

The utilization of marine microbes in the food industry arise a major question of whether or not marine microbes are food grade. A wide variety of metabolites are being produced from microbes that have potential health-promoting capabilities, but all microbes and their metabolites might not be used in food product development as they are required to comply with food regulatory requirements. However, there is no generalized definition for food-grade microorganisms, and generally, microbes with a history of safe or harmless use are regarded as food-grade microorganisms. Substantial human consumption of food over a number of generations and in a huge genetically varied population for which there exist adequate toxicological and allergenicity data to grant reasonable assurance that no harm or impairment will result from consumption of the food is defined as history of safe use (Bourdichon et al., 2012). The taxonomy, opportunistic infections, antibiotic resistance, toxic metabolites, and virulence factors and undesirable properties are the major criteria that determine the safe use of micro-organisms (Bourdichon et al., 2012). In 2002, the FAO/WHO joint committee sum-marized a few guidelines for the assessment of a microbial strain to be employed as a probiotic strain while emphasizing that genus, species, and safety evaluations including in vitro and human trials are obligatory to endorse a microbial strain as a novel pro-biotic strain (Araya et al., 2002). With recent developments in marine microbiology and perceiving the importance of having scientific evidences on microbial diversity of marine fermented food, many efforts have been made to divulge microbial communi-ties associated with traditional fermented marine foods. It was observed that a num-ber of lactic acid bacterial species, including *Bacillus*, *Micrococcus*, *Pediococcus*, and *Pseudomonas* of marine origin, were dominant in these foods.

5.5 PROTEINS AND BIOACTIVE PEPTIDES FROM MARINE MICROBES

Protein, a major source of carbon, is an essential constituent of the diet required for the survival of humans and animals. Dietary proteins have become a source of physiologi-cally active factors that have a positive effect on the functions of the body. The two most general terms in modern nutritional supplements are "bioactive proteins" and "protein hydrolysates." Proteins are large biological molecules, or macromolecules, comprising one or more long chains of amino acid residues, while peptides represent the shorter forms. Proteins present in our food can act as health promoters in two ways: First, by acting as indigestible substances in our digestive tract, they trap and expel (through feces) toxins and bile, thereby reducing the reabsorption of cholesterol from the large intestine, and second, during digestion, proteins can be converted into peptides that are absorbed into the blood circulatory system (Sohaimy, 2012). However, all the rich protein sources cannot be utilized to develop supplements without taking into consideration their bio-logical value (the amount or percentage of a protein that the body is able to absorb) (Millward et al., 2008). There are several harmful effects of meat, dairy, and soy pro-tein that are commonly used as sources of protein supplements. Moreover, besides the high cost, animal-derived proteins have encountered increasing challenges in modern nutritional market due to communicable animal diseases and preference for vegetarian diet (Fuhrman and Ferreri, 2010). In this regard, with high biological value, proteins of microbial origin could be used as an alternative source of protein for food and pharma-ceutical products (Nasseri et al., 2011).

Bioactive food proteins and peptides derived from marine microbes have been observed to play role in promoting health and lead to the reduction in the risk of disease. Much attention has been paid by consumers to natural bioactive com-pounds as functional constituents. It has been recommended that marine-derived bioactive food proteins and peptides are alternative sources for synthetic components that can contribute to consumer well-being as a part of functional foods and pharmaceuticals.

Marine organisms, especially microalgae, are regarded as a promising source of high-quality proteins. The protein content of some microalgae species, including *Chlamydomona*, *Micractinium*, *Chlorella*, *Scenedesmus*, *Spirulina*, *Dunaliella*, *Oscillatoria*, and *Euglena*, accounts for more than 50% of the dry weight, and these proteins bear high biological value (Becker, 2007). Species such as *Spirulina* and *Chlorella* with biological values of 77.6% and 71.6%, respectively, have also been considered as prospective sources of bioactive proteins. *Streptomyces platensis* is regarded as a major source of bioactive proteins in marine environments as approximately 80% of the protein nitrogen could be extracted from hexane defatted biomass. Compositional analysis of microalgae proteins revealed that this high-quality protein can be efficiently used as direct supplements or could be used for formulation of other health products such as nutraceuticals (Brown et al., 1997; Brown, 2002).

5.6 BACTERIOCINS

Majority of the Gram-positive and Gram-negative bacteria produce proteins or polypeptides, possessing antimicrobial growth inhibitory activities called bacteriocins. These are bacterial proteinaceous toxins that inhibit the growth of similar or closely related bacterial strain(s). They are phenomenologically equivalent to killing factors produced by yeast and *Paramecium*, defensins of mammals, tachyplesins of crabs, cecropins of insects, and thionins of plants but are structurally, functionally, and ecologically diverse. Although bacteriocins could be regarded as antibiotics, but unlike antibiotics that have a broad spectrum of activity, bacteriocins are observed to confine their activity to strains of species closely related to the producing species subsequently killing competitors so that the producer may flourish. In addition, bacteriocins are ribosomally synthesized peptides that are produced during the primary phase of growth, whereas antibiotics are usually secondary metabolites (Beasley and Saris, 2004). The bacteriocin-producing bacteria possess "immunity" to their own bacteriocin, allowing them, which would otherwise be susceptible, to survive and compete. Bacteriocins are low-molecular-weight cationic or amphipathic molecules and majority of them undergo posttranslational modifications. The genes encoding bacteriocins may be located on plasmids, transposons, or bacterial chromosomes and generally consist of an operon of several genes responsible for biosynthesis, posttranslational modification, and transportation, as well as the imparting immunity to the producer from its bacteriocin. Bacteriocins are considered to be safe for human consumption as these can be easily degraded by the proteases of the mammalian gastrointestinal tract. When incorporated in aqueous solutions, these molecules are usually unstructured, but when exposed to structure-promoting solvents such as trifluoroethanol or mixed with anionic phospholipid membranes, they form helical structures (Moll et al., 1999). The application of these antimicrobial compounds as a natural hurdle against pathogens and food spoilage caused by bacterial agents has been proven to be proficient (Chen and Hoover, 2003; Chopra et al., 2015a,b). Bacteriocins have gained new attention particularly in the epidemiology of nosocomial infections, which are found to be very useful in typing organisms, predominantly those that are not easy to type by regular methods (Pit, 1980; Daw et al., 1992).

5.7 PEPTIDES: SCREENING, ISOLATION, EXTRACTION, PURIFICATION, AND ANALYSIS

Considering the enormous biodiversity of marine inhabitants, the employment of appropriate strategies and methodologies that can rapidly screen various marine sources for bioactive peptides is of great interest. When designing the screening methodology, different parameters such as the probable nature of the sought-after bioactive peptide, its solubility, heat resistance, molecular weight, isoelectric point, and antimicrobial spectra have to be considered.

5.7.1 Screening of Cultures and Isolates

The initial screening for the production of bioactive peptides is usually done using the agar well diffusion technique, also known as the Kirby–Bauer well diffusion method (Chopra et al., 2014). The cell-free broth of the culture after the incubation time is tested by spreading the indicator organisms on the agar plates and then punching a well on it followed by the filling of the wells with the cell-free broth (100 µL) (Chopra et al., 2014). The plates are incubated generally overnight, and the production of the inhibitory compound is determined by observing the plates for the zone of clearance around the wells (Figure 5.1).

The size of the inhibition zone depends on how effectual the compound is at inhibiting the growth of the bacterium. A stronger compound will create a larger zone of inhibition. Another method that can be used for the initial screening is spot agar test in which the isolates are spotted onto the agar plate. After 16 h of incubation at appropriate temperature, the plates are overlaid with soft agar (0.8% agar) containing

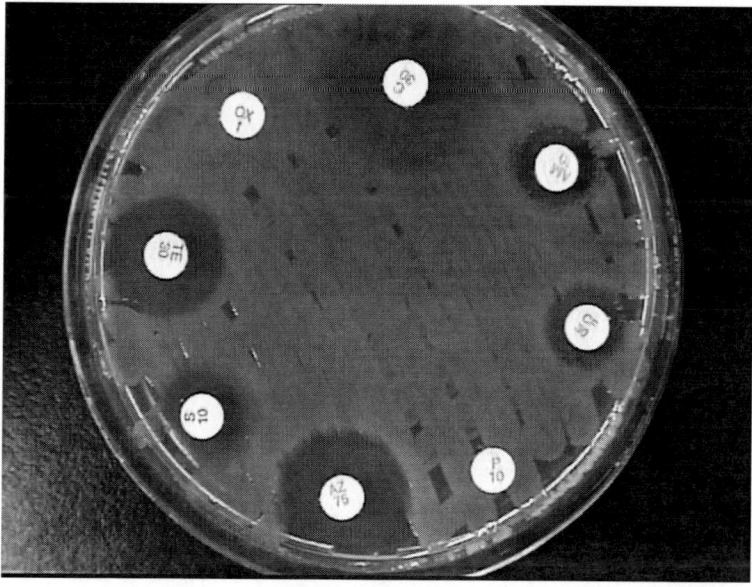

FIGURE 5.1 Zone of inhibition around the wells indicating the presence of antibacterial compounds active against *Staphylococcus aureus* (MTCC 1430).

indicator cultures. The plates are then incubated overnight to assess the activity of the isolates against each indicator strain.

5.7.2 Extraction Methods

5.7.2.1 Solvent Extraction

Various approaches for the extraction of the antimicrobial peptides have been reported. Methods of extracting peptides are based on their affinity to organic solvents, their variation in solubility in concentrated salt solutions, and their pH value, and these factors should be optimized for every peptide. The most commonly used organic solvents are ethyl acetate, methanol, chloroform, isopropanol, ether, hexane, and so on. In a typical extraction process, the cell-free broth is mixed with an organic solvent usually in the ratio of 1:3 (broth to organic solvent). The liquid is stirred vigorously followed by incubation at appropriate temperature and agitation for about 2 h. After the incubation period, the mixture is taken in a separating funnel and the liquids are allowed to get separated under gravity. The aqueous phase settles down while the organic layer or phase forms above the aqueous one. For the separation of these two phases, centrifugation at 10,000× g can also be carried out. The two phases are then collected separately and assayed for the presence of inhibitory activity. The aqueous phase is assayed as such; however, in case of the organic phase, the solvent is first evaporated by rota-vaporization, and then the sediments are dissolved in suitable buffer followed by determination of inhibitory activity. The cloudy interfacial layer formed between the two phases can also be assayed for the antimicrobial activity.

Burianek and Yousef (2000), while comparing the solvent extraction of a bacteriocin named lacidin produced by *Lactobacillus acidophilus* OSU133 using chloroform, methanol, isopropanol, and acetonitrile in 1:1 (v/v) concentration, could not detect peptide activity in the solvent phases. Chloroform-containing mixtures, however, produced precipitates at the interface layer between solvent (chloroform) and aqueous (culture supernatant) with strong bactericidal activity, and it was found to be the best solvent where maximum activity of lacidin was observed. They also tried extraction of other peptides— pediocin, nisin, and subtilin—and observed that more peptides could be extracted with chloroform. Moreover, the total activity was more in the chloroform-extracted fraction than in the nonextracted culture supernatant signifying that chloroform might have dispersed some peptide aggregates.

5.7.2.2 Ammonium Sulfate Precipitation

Salting out methods can be applied to concentrate the peptides from the culture broth. Precipitation with ammonium sulfate, most frequently used salt, is carried out by adding solid salt slowly to a protein sample at low temperature with continuous stirring until the desired saturation percentage of ammonium sulfate is reached. The solubility of proteins varies according to the ionic strength of the solution and, hence, according to the salt concentration. Two distinct effects are observed: At low salt concentrations, the solubility of the protein increases with increasing salt concentration (i.e., increasing ionic strength), an effect termed as salting in. As the salt concentration (ionic strength) is increased further, the solubility of the protein begins to decrease. At sufficiently high ionic strength, the protein will be almost completely precipitated out from the solution (salting out). However, as some peptides may precipitate at lower ammonium sulfate concentrations, or even in a small range of saturation, it is important to assay the appropriate concentration of salt that precipitates the peptide of interest. The suspension is generally incubated overnight at 4°C. Salted-out proteins are collected by centrifugation (10,000× g for 30 min) followed by the dissolution in a small volume of appropriate buffer or distilled water. This is followed by the desalting of the suspension by dialyzing with a suitable buffer or water at 4°C for 12–24 h using semipermeable membranes of appropriate molecular weight cutoff depending upon the size of the peptide of interest.

Borzenkov et al. (2014) used ammonium sulfate to precipitate the peptide secreted by *B. subtilis* BSX and found that the peptide retained its entire antibacterial activity after precipitation at 50% saturation. However, He and coworkers, while purifying bacteriocin by ammonium sulfate precipitation at 20%, 30%, 40%, 50%, 60%, 70%, and 80% saturation, found maximum inhibitory activity in the resolved precipitate with 60% saturation of ammonium sulfate. Although this procedure performed reasonably well in purifying smaller quantities of peptides from a <1.0 L bacterial culture, it has been reported to be unmanageable with large volumes in large-scale industrial production and for production of peptides for structural and functional studies (He et al., 2006). There are also reports of ammonium sulfate precipitation of peptides from culture medium resulting in variable and often low yields, in part because of much of the precipitate floating even after centrifugation and consequent difficulty in collecting these precipitates (Delves-Broughton et al., 1996; Contreras et al., 1997).

5.7.2.3 pH-Dependent Adsorption–Desorption Method

It is known that, in general, bacteriocin-producing bacterial cells adsorb the bacteriocin that they produce (Bhunia et al., 1991). If, after fermentation, the culture broth of a producer strain is adjusted to the pH for maximum adsorption of the bacteriocin onto the cell surfaces, the cells with adsorbed bacteriocin molecules could easily be removed from the culture broth by centrifugation. The peptide could then be selectively released from the cells at low pH of 1.5–2.0, and the preparation could provide large quantities of purified bacteriocin.

In a typical pH-dependent adsorption–desorption process, the culture broth after harvesting is heated to 70°C for about 30 min to make the producer cells inactive followed by the adjustment of the pH of the broth to pH 6.5 and incubation at 30°C and 100 rpm for 2–4 h to allow the peptide molecules to get adsorb on the producer cells. After that, the cells are harvested by centrifugation and washed with buffer at the same pH, that is, pH 6.5. The elution of the peptide molecules is carried out by adjusting the pH to lower values, pH 1–2.5, using HCl. The eluted peptides are then assayed for inhibitory activity after the adjustment of pH to 7.

5.7.2.4 Extraction at Low pH

The peptides can also be precipitated out of the culture broth by adjusting the pH of the broth to pH 2.0 by adding 4 M HCl and then keeping at 4°C overnight (Borzenkov et al., 2014). The precipitated proteins are separated by centrifugation at 10,000× g for 20 min at 4°C, and the pellet is resolved in water to which isopropanol is then added and mixed. The resultant suspension is kept at 4°C for 1 h with shaking and centrifuged (12,800× g) for 10 min. This was followed by removal of the pellet and evaporation of the residual isopropanol in a rotor evaporator at 60°C. The pellet is then assayed for the presence of inhibitory activity.

5.7.2.5 Hydrophobic Resins

For the selective separation of peptides from the culture broth, hydrophobic resins such as Amberlite XAD-16 (Tulini et al., 2014) and Diaion HP-20 (Chopra et al., 2014) are used. As compared to ammonium sulfate precipitation that precipitates all the proteins, few studies have reported utilization of hydrophobic interaction of resins to extract selective peptides (Sebei et al., 2007; Appleyard et al., 2009). The peptides bind to the resins packed in a column by passing cell-free broth through it. This is followed by washing of the resins, generally, with 10%–30% of organic solvents like methanol, isopropanol, and acetonitrile. The bound peptides are then eluted with 100% of the organic solvent followed by the evaporation of the solvent using a Rotavapor. The residue left behind is then resolved into a suitable buffer and assayed for the inhibitory activity.

5.7.3 Purification of Peptides

5.7.3.1 Ion Exchange Chromatography

All the methods described earlier result in the extraction of peptides in a crude form, which need to be purified further using various chromatographic techniques. Again, the chromatographic procedure to be followed depends upon the nature of the peptide of interest.

Peptides contain charged groups on their surfaces that enhance their interactions with solvent water and hence their solubility. Charged residues can be cationic or anionic, and it is noteworthy that even polar residues can also be charged under certain pH conditions. Types of ion exchange adsorbents are given in Table 5.1 and Figure 5.2.

As the peptides have unique amino acid sequences, the net charge on a peptide at physiological pH is determined ultimately by the balance between these charges. In general, these peptides are cationic in nature and therefore cation ion exchange would be the method of choice. The basis for ion exchange chromatography is the attraction between peptides in solution and the charged groups on the ion exchanger. The strength of the interaction depends upon the charge of the peptides and the ion exchangers, the dielectric constant of the medium, and the competition from other ions for the charged groups of the ion exchanger and peptide. When the concentration of the competing ions is low, the peptide of interest adsorbs to the ion exchanger, and at high competing ion concentration, the peptide desorbs. pH is one of the most important parameter affecting peptide binding, as it determines the charge on both the peptide and the ion exchanger. The net charge of the peptide varies with pH, and the binding should occur when the net charge

TABLE 5.1   Types of Ion Exchange Adsorbents

Ion Exchange Adsorbent Type	Cation Exchanger	Anion Exchanger
Charge of functional groups of ion exchanger	Negative	Positive
Net charge of proteins of interest	Positive (protein is a cation)	Negative (protein is an anion)
Advised buffer pH	<pI – 1	>pI + 1
Resulting pH in the vicinity of the solid phase	Up to 1 unit lower than buffer pH	Up to 1 unit higher than buffer pH

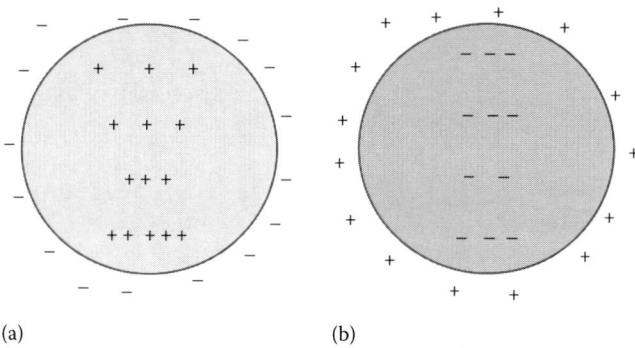

(a)   (b)

FIGURE 5.2   Types of ion exchangers: (a) Anion exchanger and (b) cation exchanger.

is of opposite sign to that of the ion exchanger. At the pH values far away from the pI, peptides bind strongly and do not desorb with low ionic strength buffers.

The most common functional groups of the cationic exchangers and their pK_a values are provided in Table 5.2.

The buffer to be used for the cation ion exchange chromatography plays an essential role in the adsorption and subsequent desorption of the peptides. In general, buffers with the following characteristics should be used (Karlsson and Hirsh, 2011):

- Adequate buffering capacity
- Ions that assist good selectivity
- Salts that are easily removed

The buffers with high capacity and with buffering ions that do not bind to the ion exchanger are recommended. For a cation exchanger, negatively charged buffering ions such as phosphate, carbonate, acetate, or morpholino-ethanesulfonate can be used with the counterions Na^+ or K^+. The ion exchange resins are selectively eluted by slowly increasing the ionic strength (this disrupts ionic interactions between the peptide and column matrix competitively) or by altering the pH (the reactive groups on the peptides lose their charge) (Dolman et al., 2002). Chopra et al. (2014) reported the purification of the peptide from marine isolate *B. sonorensis* MT93 using SP sepharose resin at pH 7.2. Although the majority of the peptides obtained from marine organisms are cationic in nature, few are reported to be anionic and the purification of such peptides is carried out

TABLE 5.2   Functional Groups of the Cationic Exchangers

Type	Name	Designation	pK_a	Structure
	Methacrylate		6.5	$-CH_2CH(CH_3)COOH$
Strong	Sulfonate	S	2	$-OCH_2SO_3H$
	Orthophosphate	P	3 and 6	$-OPO_3H_2$
Strong	Sulfopropyl	SP	2–2.5	$-OCH_2CH_2CH_2SO_3H$
Strong	Sulfoxyethyl	SE	2	$-OCH_2CH_2SO_3H$
Weak	Carboxymethyl	CM	3.5–4	$-OCH_2COOH$

TABLE 5.3   Functional Groups of the Anionic Exchangers

Type	Name	Designation	pK_a	Structure
Strong	Trimethyl hydroxypropyl	QA		$-OCH_2CH(OH)NH(C_2H_5)_2$
Strong	Quaternary amino ethyl	QAE		$-OCH_2CH_2N^+(C_2H_5)_2CH_2CH(OH)CH_3$
Strong	Quaternary amine	Q		$-OCH_2N^+(CH_3)_3$
Strong	Triethylamine	TEAE	9.5	$-OCH_2N^+(C_2H_5)_3$
Weak	Dimethylaminoethyl	DMAE	10	$-OCH_2CH_2NH(CH_3)_2$
Strong	Trimethylaminoethyl	TMAE		$-OCH_2\ CH_2N^+(CH_3)_3$
Weak	Diethylaminoethyl	DEAE	6–9	$-OCH_2NH(C_2H_5)_2$

using anion exchanger resins. The common functional groups of anion exchangers and their pK_a values are summarized in Table 5.3.

In anion exchange chromatography, lowering the pH of the mobile phase buffer will cause the molecule to become more protonated and hence more positively (and less negatively) charged. The result is that the protein will no longer be able to form ionic interaction with the positively charged solid support, which may cause the molecule to elute from the column. Rajaram et al. (2010) purified a bacteriocin produced by marine isolate *L. lactis* using DEAE-cellulose at pH 7.0. After washing, the bound proteins were eluted stepwise using buffers of increasing molarity and decreasing pH values at room temperature (approx. 25°C).

5.7.3.2 Gel Permeation

The chromatographic step that is used for further purification of peptides is gel permeation or size exclusion chromatography that separates peptides on the basis of their size or hydrodynamic volume (radius of gyration) (Figure 5.3). This differs from other separation techniques that depend upon chemical or physical interactions to separate peptides; it involves the use of porous beads packed in a column. The smaller peptides can enter the pores more easily and therefore spend more time in these pores, increasing their retention time. Conversely, larger peptides spend little, if any, time in the pores and are eluted rapidly.

There are a variety of chromatography matrices for a range of molecular weights that can be separated. As the peptides are generally smaller in size, so is the matrix or column to be used for the separation and better resolution depends upon the size of the peptide. For example, Longeon et al. (2004) purified an antibacterial peptide from marine *Pseudoalteromonas* sp. strain X 153 using Superdex 200 HR 10/30 (Pharmacia) equilibrated with 25 mM ammonium bicarbonate and 0.15 M NaCl and eluted with the same solvent at a flow rate of 0.4 mL/min. Table 5.4 summarizes the various matrix or prepacked columns used for the purification of bioactive peptides.

5.7.3.3 Reverse-Phase HPLC

The final step in the purification of bioactive compounds is reverse-phase high-performance liquid chromatography (RP-HPLC). Since the peptides are hydrophobic in nature, the separation is based upon the hydrophobicity. It is the opposite of normal-phase chromatography and results from the adsorption of hydrophobic molecules onto a

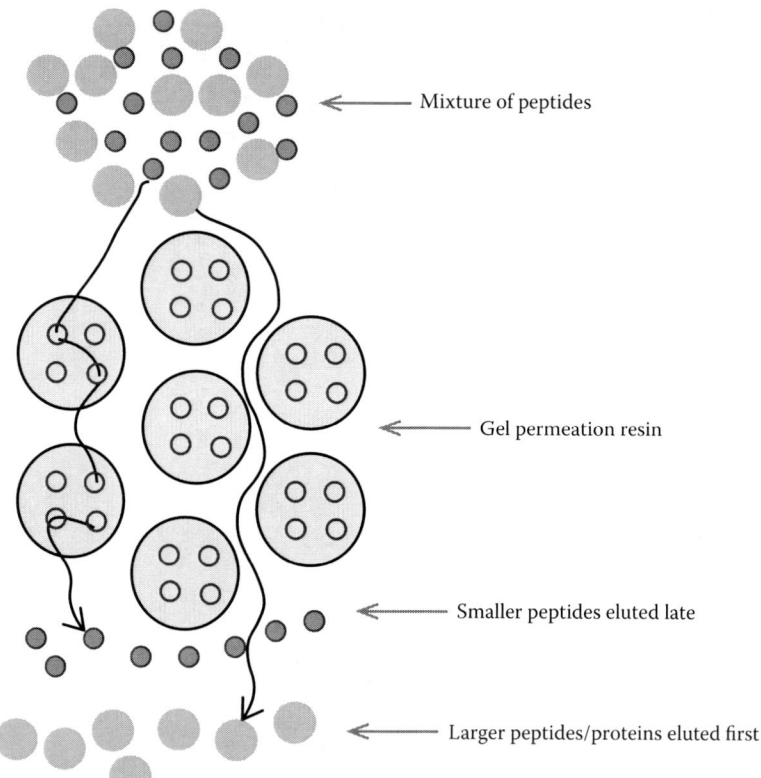

Mixture of peptides

Gel permeation resin

Smaller peptides eluted late

Larger peptides/proteins eluted first

FIGURE 5.3   Overview of gel filtration chromatography.

hydrophobic solid support in a polar mobile phase. Decreasing the mobile phase polarity by adding more organic solvent reduces the hydrophobic interaction between the solute and the solid support resulting in desorption.

The more hydrophobic the molecule, the more time it will spend on the solid support and the higher the concentration of the organic solvent that is required to promote desorption. The most commonly used matrices for RP-HPLC are summarized in Table 5.5. Mixtures of water or aqueous buffers and organic solvents are used for elution of peptides. The most common water-miscible organic solvents used are acetonitrile, methanol, and tetrahydrofuran. Other solvents can be used such as ethanol or 2-propanol (isopropyl alcohol). Elution can be performed isocratically (the water-solvent composition does not change during the separation process) or by using a solution gradient (the water-solvent composition changes during the separation process, usually by decreasing the polarity). The pH of the mobile phase can have a significant role on the retention of peptides and can change the selectivity of certain peptides. A general way to increase the hydrophobicity of charged components, enhance binding to the medium, and so alter retention time is to add ion-pairing agents such as trifluoroacetic acid (TFA) to the eluent. An et al. (2015) reported the purification of a bacteriocin named as CAMT2 produced by *B. amyloliquefaciens* using reverse-phase C18 column using solvent A (60% methanol) and solvent B (40% water).

TABLE 5.4   Matrix/Prepacked Columns Used for the Purification of Bioactive Peptides

Media	Type	Particle Size of Hydrated Beads (μm)	Fractionation Range (kDa)	pH Stability	Packaging
[a]Sephadex™				2–10	Bulk media
G25	Dextran	20–100	0.1–5		
G50		20–100	0.5–30		
[b]Bio-gel™				2–10	Prepacked
P-2			0.1–1.8		
P-4	Polyacrylamide		0.8–4.0		
P-6			1.0–6.0		
P-10			1.5–20		
[a]Sephacryl™ S-100 HR	Dextran/ bisacrylamide	25–75	1–100	2–11	Prepacked
[a]Superdex™				1–14	
Peptide	Agarose	13–34	0.1–7.0		Prepacked
30	Dextran	34	<10		Bulk media
75		13–34	3.0–70.0		Prepacked
UltrogelAcA 202	Agarose/acrylamide				Prepacked
		60–140	1–15	3–10	

Manufacturers: [a]GE Healthcare (Uppsala, Sweden); [b]Bio-Rad Laboratories (Hercules, CA, USA).

TABLE 5.5   Matrices for RP HPLC of Peptides

Packing	Material Type	Pore Size	Particle Size (μm)
[a]Luna™	Silica C8, C18	100 A°	3–5
[a]Jupiter™	Silica C4, C18	300 A°	3–10
[b]Kromasil	Silica C4, C8, C18	100 A°	3.5–7
[c]Source™ 15RPC	Polystyrene/divinylbenzene	Proprietary	15
[d]Vydac TP	Spheroidal silica C4, C8, C18	300 A°	10–30
[e]Zorbax 300 Stable Bond	Silica C4, C8, C18	300 A°	10–20

Manufacturers: [a]Phenomenex; [b]AkzoNobel; [c]GE Healthcare; [d]Grace; [e]Agilent.

5.7.4 Tricine Polyacrylamide Gel Electrophoresis (Tricine-SDS-PAGE)

Tricine-SDS-PAGE is an efficient way of separating low-molecular-mass peptides. Usually, a 16% gel (with pH 8.45) is employed to separate small peptides using two running buffers: a cathode buffer (100 mM Tris, 100 mM tricine, and 0.1% [w/v] SDS, pH 8.25) and an anode buffer (100 mM Tris, pH 8.9). However, to obtain highly resolved bands, this method requires three gels (stacking, spacer, and resolving gel) (Schagger and von Jagow, 1987; Schagger, 2006) and an addition of urea (Okajima et al., 1993) in the resolving gel. The stacking of small proteins in the presence of high concentration of SDS as in case of glycine-SDS-PAGE is difficult because small peptides form complexes of protein and detergent of the "same" size and charge as the SDS micelle itself (Fish et al., 1970). Tricine, used as a trailing ion, allows a better resolution of small proteins than in glycine-SDS-PAGE systems. At usual pH values between 6.8 and 8.8, tricine migrates much faster than glycine in a stacking gel despite its higher molecular mass as it has a higher negative (more negative) charge than glycine, thus allowing it to migrate faster. In addition, its high ionic strength causes more ion movement and less protein movement. This allows for low-molecular-weight peptides to be separated in lower percent acrylamide gels.

5.7.5 Zymography

To determine the in situ antibacterial activity of the peptides, after running the tricine-SDS-PAGE, the gel is cut into two parts. One part is stained with Coomassie Brilliant Blue G-250 after fixing with fixing solution (10% acetic acid, 50% methanol, 40% water). The second half of the gel is used for in situ detection of bioactivity upon fixing it in a mixture of 2-propanol, acetic acid, and water (25:10:65) for 15 min and washing with sterile water repeatedly. The gel is then placed on a prepoured agar spread with an indicator organism, and upon incubation of the plate, the zone of growth inhibition could be observed (Chopra et al., 2014).

5.8 QUANTIFICATION AND DETERMINATION OF COMPOSITION OF BIOACTIVE PEPTIDES

Development of functional foods that contain functional peptides requires not only identification of the active form, but also quantitative analysis and study of the peptide stability in complex food and biological matrices. Generally, identification of a peptide necessitates its purification and structural analysis. However, in most cases they are time-consuming and accompanied by some difficulties. In addition, rapid detection and identification of peptides in the early stage of screening are advantageous in order to avoid the risk of discovering a peptide identical to a known one. Mass spectrometry (MS) has become the analytical method of choice, offering selectivity and sensitivity for peptide identification, characterization, and quantification. Identification of bioactive peptides is usually done by desorption techniques such as plasma desorption, secondary ion MS, and fast atom bombardment (FAB)-MS. FAB ionization is one of the pioneering desorption techniques that has contributed to the substantial progress in the mass determination of peptides. Other methods that have been employed for the identification and quantification of peptides include methods utilizing triple quadrupole ion trap, quadrupole time of flight (Q-Tof), and electrospray ionization (ESI)-MS. These methods not only help in the detection of the known bioactive peptides but also in the identification of novel peptides by accurate mass determination. Direct supernatant of the culture can be used for the identification. But for more specificity, the purified peptides are run on tricine-PAGE prior to the identification by these techniques. The usual method of peptide sample preparation involves the excision of bands from tricine-PAGE gels followed by the reduction with dithiothreitol to break disulfide linkages, alkylation with iodoacetamide, and then digestion with one of the protease like trypsin, chymotrypsin, cyanogen bromide, and proteinase K. The resultant peptides are extracted in washes of ammonium bicarbonate solution, acetonitrile, and 1.0% formic acid. The extraction solvent is then removed under vacuum and the peptides are resuspended in water with 0.1% TFA.

A full structural characterization of the peptides in the samples requires the combination of various MS techniques. Sometimes the results of MALDI measurements compared to those of ESI are remarkably different, so that only a combination of both ionization techniques allows a comprehensive analysis of the peptide samples (Stapels et al., 2004). In addition to the task of identification of bioactive peptides, MALDI-TOF-MS has been used to better understand the digestion processes that lead to the formation of bioactive peptides. The raw MS/MS data are searched against NCBI entries of bacteria (eubacteria) (NIH, Bethesda, Maryland) and Swiss-Prot database using the Mascot algorithm (Matrix Science). Bioactive peptide–specific databases like Antimicrobial Peptide Database, Bactibase, Prowl, and Bagel are also used for the sequence alignment with already identified peptides. Chen et al. (2010) identified two antifungal lipopeptides produced by isolate, from deep sea sediment, *B. amyloliquefaciens* SH-B10 using tandem Q-TOF MS as C16 fengycin and a new fengycin with an aminobutyric acid at the six position of the peptide backbone.

5.9 CONCLUSIONS

In recent years, there has been an enormous progress in the study of functional foods and nutraceuticals and their importance to the health of humans and animals, but there is a need for further research in many areas of this ancient, yet modern, field. To begin with, there is ample scope for more basic research that will lead to a better understanding of the mechanisms whereby functional foods and nutraceuticals exert their beneficial effects. This, in turn, should lead to advances in the applied aspects of the field. Thus, the biologically active components must be identified and quantified using rigorous, standardized, internationally accepted methods or protocols. Another important research area could be the exploration of unknown medicinally beneficial substances in marine microorganisms. It is essential, however, that the discovery and subsequent exploitation of new sources of functional foods and nutraceuticals in nature should be accomplished without damaging the environment.

REFERENCES

An, J., Wenjuan, Z., Ying, L. et al. Purification and characterization of a novel bacteriocin Camt2 produced by *Bacillus amyloliquefaciens* isolated from marine fish *Epinephelus areolatus*. *Food Control*, 51 (2015): 278–282.

Appleyard, A.N., Shaila, C., Daniel, M.R. et al. Dissecting structural and functional diversity of the *Lantibiotic mersacidin*. *Chem. Biol.*, 16(5) (2009): 490–498.

Araya, M., Morelli, L., Reid, G. et al. Guidelines for the evaluation of probiotics in food. Report of a Joint FAO/WHO working group on drafting guidelines for the evaluation of probiotics in food, London, Ontario, Canada, 2002.

Beasley, S.S. and Saris, P.E.J. Nisin-producing *Lactococcus lactis* strains isolated from human milk. *Appl. Environ. Microbiol.*, 70(8) (2004): 5051–5053.

Becker, E.W. Micro-algae as a source of protein. *Biotechnol. Adv.*, 25(2) (2007): 207–210.

Bernan, V.S., Greenstein, M., and Carter, G.T. Mining marine organisms as a source of new antimicrobials and antifungals. *Curr. Med. Chem. Anti-Infect. Agents*, 3 (2004): 181–195.

Bernan, V.S., Greenstein, M., and Maiese, W.M. Marine microorganisms as a source of new natural products. *Adv. Appl. Microbiol.*, 43 (1997): 57–90.

Bertoldo, C. and Antranikian, G. Starch-hydrolyzing enzymes from *Thermophilic archaea* and bacteria. *Curr. Opin. Chem. Biol.*, 6(2) (2002): 151–160.

Bhakuni, D.S. and Rawat, D.S. *Bioactive Marine Natural Products*. Anamaya Publishers, New Delhi, India, 2005.

Bhatnagar, I. and Se-Kwon, K. Immense essence of excellence: Marine microbial bioactive compounds. *Mar. Drugs*, 8(10) (2010): 2673–2701.

Bhunia, A.K., Johnson, M.C., Ray, B., and Kalchayanand, N. Mode of action of pediocin AcH from *Pediococcus acidilactici*-H on sensitive bacterial strains. *J. Appl. Bacteriol.*, 70(1) (1991): 25–33.

Bizani, D., Motta, A.S., Morrissy, J.A.C., Terra, R.M.S., Souto, A.A., and Brandelli, A. Antibacterial activity of cerein 8a, a bacteriocin-like peptide produced by *Bacillus cereus*. *Int. Microbiol.*, 8(2) (2005): 125–131.

Blunt, J.W., Copp, B.R., Munro, M.H.G., Northcote, P.T., and Prinsep, M.R. Marine natural products. *Nat. Prod. Rep.*, 21(1) (2004): 1–49.

Borzenkov, V., Surovtsev, V., and Dyatlov, I. Obtaining bacteriocins by chromatographic methods. *Adv. Biosci. Biotechnol.*, 5 (2014): 446–451.

Bourdichon, F., Berger, B., Casaregola, S. et al. Safety demonstration of microbial food cultures (MFC) in fermented food products. *Bull. Int. Dairy Fed.*, 455 (2012): 1–62.

Brown, M.R. Nutritional value and use of microalgae in aquaculture. In: *Avances en Nutricion Acuicola VI*. Memorias del VI Simposium Internacional de Nutricion Acuicola, Walker, J.M., Ed. Humana Press, Totowa, NJ, Vol. 3, 2002, pp. 281–292.

Brown, M.R., Jeffrey, S.W., Volkman, J.K., and Dunstan, G.A. Nutritional properties of microalgae for mariculture. *Aquaculture*, 151(1–4) (1997): 315–331.

Buck, J.D., Meyers, S.P., and Kamp, K.M. Marine bacteria with antiyeast activity. *Science*, 138(3547) (1962): 1339–1340.

Burianek, L.L. and Yousef, A.E. Solvent extraction of bacteriocins from liquid cultures. *Lett. Appl. Microbiol.*, 31(3) (2000): 193–197.

Chang, L.C., Whittaker, N.F., and Bewley, C.A. Crambescidin 826 and dehydrocrambine A: New polycyclic guanidine alkaloids from the marine sponge Monanchora sp that inhibit HIV-1 fusion. *J. Nat. Products*, 66(11) (2003): 1490–1494.

Chen, H. and Hoover, D.G. Bacteriocins and their food applications. *Compr. rev. food sci. food saf.*, 2 (2003): 82–100.

Chen, L., Nan, W., Xuemei, W., Jiangchun, H., and Shujin, W. Characterization of two anti-fungal lipopeptides produced by *Bacillus amyloliquefaciens* Sh-B10. *Bioresour. Technol.*, 101(22) (2010): 8822–8827.

Cheng, X.C., Varoglu, M., Abrell, L., Crews, P., Lobkovsky, E., and Clardy, J. Chloriolins A-C, chlorinated sesquiterpenes produced by fungal cultures separated from a Jaspis marine sponge. *J. Org. Chem.*, 59(21) (1994): 6344–6348.

Chopra, L., Gurdeep, S., Kautilya, K.J., and Debendra, K.S. Sonorensin: A new bacteriocin with potential of an anti-biofilm agent and a food biopreservative. *Sci. Rep.*, 5 (2015b): 13412.

Chopra, L., Gurdeep, S., Kautilya, K.J., Himanshu, V., and Debendra, K.S. Bioprocess development for the production of sonorensin by *Bacillus sonorensis* Mt93 and its application as a food preservative. *Bioresour. Technol.*, 175 (2015a): 358–366.

Chopra, L., Gurdeep, S., Vikas, C., and Debendra, K.S. Sonorensin: An antimicrobial peptide, belonging to the heterocycloanthracin subfamily of bacteriocins, from a new marine isolate, *Bacillus sonorensis* Mt93. *Appl. Environ. Microbiol.*, 80(10) (2014): 2981–2990.

Contreras, B.G.L., DeVuyst, L., Devreese, B., Busanyova, K., Raymaeckers, J., Bosman, F., Sablon, E., and Vandamme, E.J. Isolation, purification, and amino acid sequence of lactobin a, one of the two bacteriocins produced by *Lactobacillus amylovorus* LMG P-13139. *Appl. Environ. Microbiol.*, 63(1) (1997): 13–20.

Crawford, K., Heatley, N.G., Boyd, P.F., Hale, C.W., Kelley, B.K., Miller, G.A., and Smith, N. Antibiotic production by a species of cephalosporium. *J. Gen. Microbiol.*, 6(1–2) (1952): 47–59.

Davidson, B.S. New dimensions in natural-products research—Cultured marine microorganisms. *Curr. Opin. Biotechnol.*, 6(3) (1995): 284–291.

Daw, M.A., Corcoran, G.D., Falkiner, F.R., and Keane, C.T. Application and assessment of cloacin typing of *Enterobacter cloacae*. *J. Hosp. Infect.*, 20(3) (1992): 141–151.

DeLong, E.F. Modern microbial seascapes. *Nat. Rev. Microbiol.*, 5(10) (2007): 755–757.

Delves-Broughton, J., Blackburn, P., Evans, R. J., and Hugenholtz, J. Applications of the bacteriocin, nisin. *Antonie Van Leeuwenhoek*, 69 (1996): 193–202.

Dolman, C., Page, M., and Thorpe, R. Purification of IgG using ion-exchange HPLC. In: *The Protein Protocols Handbook* (2nd Edition). Walker, J.M., Ed. Humana Press, Totowa, NJ, 2002, pp. 989–990.

Etahiri, S., Bultel-Ponce, V., Caux, C., and Guyot, M. New bromoditerpenes from the red alga *Sphaerococcus coronopifolius*. *J. Nat. Products*, 64(8) (2001): 1024–1027.

Fenical, W. Chemical studies of marine-bacteria—Developing a new resource. *Chem. Rev.*, 93(5) (1993): 1673–1683.

Fennell, B.J., Carolan, S., Pettit, G.R., and Bell, A. Effects of the antimitotic natural product dolastatin 10, and related peptides, on the human malarial parasite *Plasmodium falciparum*. *J. Antimicrob. Chemother.*, 51(4) (2003): 833–841.

Fish, W.W., Reynolds, J.A., and Tanford, C. Gel chromatography of proteins in denaturing solvents. Comparison between sodium dodecyl sulfate and guanidine hydrochloride as denaturants. *J. Biol. Chem.*, 245(19) (1970): 5166–5168.

Food and Agriculture Organization of the United Nations (FAO). Report on Functional Foods, Food Quality and Standards Service (AGNS), 2007. http://www.fao.org/ag/agn/agns/files/Functional_Foods_Report_Nov2007.pdf. Accessed on February 25, 2010.

Fuesetani, N. In *Drugs from the Sea*. Karger Publishers, Basel, Switzerland, 2000, pp. 1–5.

Fuhrman, J. and Deana M.F. Fueling the vegetarian (vegan) athlete. *Curr. Sports Med. Rep.*, 9(4) (2010): 233–241.

Gopal, R., Vijayakumaran, M., Vennkatesan, R., and Kathiroli, S. Marine organisms in Indian medicine and their future prospects. *Nat. Prod. Radiance*, 7(2) (2008): 139–145.

Granum, P.E. *Bacillus cereus*. In: *Food Microbiology. Fundamentals and Frontiers*. Doyle, M.P., and Buchet, L.R., Eds. ASM Press, Washington, DC, 2007, pp. 445–455.

He, L., Weiliang, C., and Yang, L. Production and partial characterization of bacteriocin-like pepitdes by *Bacillus licheniformis* Zju12. *Microbiol. Res.*, 161(4) (2006): 321–326.

Hong, H.A., Duc, L.H., and Cutting, S.M. The use of bacterial spore formers as probiotics. *Fems Microbiol. Rev.*, 29(4) (2005): 813–835.

Hosoi, T. and Kiuchi, K. Natto—A food made by fermenting cooked soybeans with *Bacillus subtilis* (natto). In: *Handbook of Fermented Functional Foods*. Farnworth, E.R., Ed. CRC Press, Boca Raton, FL, 2003, pp. 227–245.

Hugenholtz, P. and Pace, N.R. Identifying microbial diversity in the natural environment: A molecular phylogenetic approach. *Trends Biotechnol.*, 14(6) (1996): 190–197.

Imhoff, J.F., Antje, L., and Jutta, W. Bio-mining the microbial treasures of the ocean: New natural products. *Biotechnol. Adv.*, 29(5) (2011): 468–482.

Jackson, C-J.C. and Paliyath, G. Functional foods and nutraceuticals. In: *Functional Foods, Nutraceuticals and Degenerative Disease Prevention*. Paliyath, G., Bakovic, M., and Shetty, K., Eds. Wiley-Blackwell, Oxford, UK, 2011, pp. 11–43.

Jay, M.J., Loessner, J.M., Golden, D.A. *Modern Food Microbiology* (7th Edition). Springer, New York, 2005.

Joffe, S. and Thomas, R. Phytochemicals: A renewable global resource. *Biotech. News Inform.*, 1 (1989): 697–700.

Kakeya, H., Takahashi, I., Okada, G., Isono, K., and Osada, H. Epolactaene, a novel neuritogenic compound human neuroblastoma-cells, produced by a marine fungus. *J. Antibiot.*, 48(7) (1995): 733–735.

Karlsson, E. and Hirsh, I. Ion exchange chromatography. In: *Protein Purification: Principles, High Resolution Methods, and Applications*. Janson, J.C., Ed. John Wiley & Sons, Hoboken, NJ, 2011, pp.93–134.

Khudyakova, Y.V., Piukin, M.V., Smetina, O.P., Slinkina, N.A., and Aleshko, S.N. *Marine Fungi as Potential Producers of Antibiotics*. Pacific Institute of Bioorganic chemistry of Far East Academy of Sciences, Vladivostok, Russia, 2004, p. 185.

Kobayashi, H., Namikoshi, M., Yoshimoto, T., and Yokochi, T. A screening method for antimitotic and antifungal substances using conidia of *Pyricularia oryzae*, modification and application to tropical marine fungi. *J. Antibiot.*, 49(9) (1996): 873–879.

Kobayashi, J. and Ishibashi, M. Bioactive metabolites of symbiotic marine microorganisms. *Chem. Rev.*, 93(5) (1993): 1753–1769.

Konig, G.M., Wright, A.D., and Franzblau, S.G. Assessment of antimycobacterial activity of a series of mainly marine derived natural products. *Planta Med.*, 66(4) (2000): 337–342.

Kuniyoshi, M., Wahome, P.G., Miono, T., Hashimoto, T., Yokoyama, M., Shrestha, K.L., and Higa, T. Terpenoids from *Laurencia luzonensis. J. Nat. Products*, 68(9) (2005): 1314–1317.

Llagostera, E., Soto-Cerrato, V., Montaner, B., and Perez-Tomas, R. Prodigiosin induces apoptosis by acting on mitochondria in human lung cancer cells. *Ann. N Y Acad. Sci.*, 1010 (2003): 178–181.

Lodemann, U., Lorenz, B.M., Weyrauch, K.D., and Martens, H. Effects of *Bacillus cereus* var. toyoi as probiotic feed supplement on intestinal transport and barrier function in piglets. *Arch. Anim. Nutr.*, 62(2) (2008): 87–106.

Longeon, A., Peduzzi, J., Barthelemy, M., Corre, S., Nicolas, J.L., and Guyot, M. Purification and partial identification of novel antimicrobial protein from marine *Bacterium pseudoalteromonas* species strain X 4 53. *Mar. Biotechnol.*, 6(6) (2004): 633–641.

Luescher-Mattli, M. Algae, a possible source for new drugs in the treatment of HIV and other viral diseases. *Curr. Med. Chem.*, 2 (2003): 219–225.

Manoharachary, C., Sridhar, K., Singh, R., Adholeya, A., Suryanarayanan, T.S., Rawat, S., and Johri, B.N. Fungal biodiversity: Distribution, conservation and prospecting of fungi from India. *Curr. Sci.*, 89(1) (2005): 58–71.

Maskey, R.P., Helmke, E., Kayser, O., Fiebig, H.H., Maier, A., Busche, A., and Laatsch, H. Anti-cancer and antibacterial trioxacarcins with high anti-malaria activity from—A marine streptomycete and their absolute stereochemistry. *J. Antibiot.*, 57(12) (2004): 771–779.

Millward, D.J., Layman, D.K., Tome, D., and Schaafsma, G. Protein quality assessment: Impact of expanding understanding of protein and amino acid needs for optimal health. *Am. J. Clin. Nutr.*, 87(5) (2008): 1576S–1581S.

Moll, G.N., Konings, W.N., and Driessen, A.J.M. Bacteriocins: Mechanism of membrane insertion and pore formation. *Antonie Van Leeuwenhoek Int. J. Gen. Mol. Microbiol.*, 76(1–4) (1999): 185–198.

Montaser, R. and Hendrik, L. Marine natural products: A new wave of drugs?. *Future Med. Chem.*, 3(12) (2011): 1475–1489.

Namikoshi, M., Akano, K., Meguro, S., Kasuga, I., Mine, Y., Takahashi, T., and Kobayashi, H. A new macrocyclic trichothecene, 12,13-deoxyroridin E, produced by the marine-derived fungus *Myrothecium roridum* collected in Palau. *J. Nat. Products*, 64(3) (2001): 396–398.

Nasseri, A., Rasoul-Amini, S., Morowvat, M., and Ghasemi, Y. Single cell protein: Production and process. *Am. J. Food Technol.*, 6(2) (2011): 103–116.

Okajima, T., Tanabe, T., and Yasuda, T. Nonurea sodium dodecyl-sulfate polyacrylamide-gel electrophoresis with high-molarity buffers for the separation of proteins and peptides. *Anal. Biochem.*, 211(2) (1993): 293–300.

Pandey, N., Meena, R.P., Rai, S.K., and Pandey-Rai, S. Medicinal plants derived neutraceuticals: A re-emerging health aid. *Int. J. Pharma BioSci.*, 2(4) (2011): 419–441.

Pathirana, C., Jensen, P.R., and Fenical, W. Marinone and debromomarinone—Antibiotic sesquiterpenoid naphthoquinones of a new structure class from a marine bacterium. *Tetrahedron Lett.*, 33(50) (1992): 7663–7666.

Perez-Tomas, R., Montaner, B., Llagostera, R., and Soto-Cerrato, V. The prodigiosins, proapoptotic drugs with anticancer properties. *Biochem. Pharmacol.*, 66(8) (2003): 1447–1452.

Pitt, T.L. State of the art: Typing *Pseudomonas aeruginosa*. *J. Hosp. Infect.*, 1(3) (1980): 193–199.

Rajaram, G., Manivasagan, P., Thilagavathi, B., and Saravanakumar, A. Purification and characterization of a bacteriocin produced by *Lactobacillus lactis* isolated from marine environment. *Adv. J. Food Sci. Technol.*, 2(2) (2010): 138–144.

Schagger, H. Tricine-SDS-Page. *Nat. Protoc.*, 1(1) (2006): 16–22.

Schagger, H. and Vonjagow, G. Tricine sodium dodecyl-sulfate polyacrylamide-gel electrophoresis for the separation of proteins in the range from 1-kda to 100-kda. *Anal. Biochem.*, 166(2) (1987): 368–379.

Sebei, S., Zendo, T., Boudabous, A., Nakayama, J., and Sonomoto, K. Characterization, N-terminal sequencing and classification of cerein MRX1, a novel bacteriocin purified from a newly isolated bacterium: *Bacillus cereus* MRX1. *J. Appl. Microbiol.*, 103(5) (2007): 1621–1631.

Shin, J., and Fenical, W. Isolation of gliovictin from the marine deuteromycete *Asteromyces cruciatus*. *Phytochem.*, 26(12) (1987): 33–47.

Sobhy, S.E. Functional foods and nutraceuticals-modern approach to food science. *World Appl. Sci. J.*, 20(5) (2012): 691–708.

Stapels, M.D., Cho, J.C., Giovannoni, S.J., and Barofsky, D.F. Proteomic analysis of novel marine bacteria using MALDI and ESI mass spectrometry. *J. Biomol. Tech.*, 15(3) (2004): 191–198.

Stein, T., Heinzmann, S., Dusterhus, S., Borchert, S., and Entian, K.D. Expression and functional analysis of the subtilin immunity genes spaifeg in the subtilin-sensitive host *Bacillus subtilis* MO1099. *J. Bacteriol.*, 187(3) (2005): 822–828.

Trischman, J.A., Jensen, P.R., and Fenical, W. Halobacillin—A cytotoxic cyclic acyl-peptide of the iturin class produced by a marine bacillus. *Tetrahedron Lett.*, 35(31) (1994): 5571–5574.

Tulini, F.L., Lohans, C.T., Bordon, K.C.F., Zheng, J., Arantes, E.C., Vederas, J.C., and De Martinis, E.C.P. Purification and characterization of antimicrobial peptides from fish isolate *Carnobacterium maltaromaticum* C2: Carnobacteriocin X and carnolysins A1 and A2. *Int. J. Food Microbiol.*, 173 (2014): 81–88.

Venkateshwar Goud, T., Srinivasa Reddy, N., Raghavendra Swamy, N., Siva Ram, T., and Venkateswarlu, Y. Anti-HIV active petrosins from the marine sponge *Petrosia similis*. *Biol. Pharm. Bull.*, 26(10) (2003): 1498–1501.

Waters, A.L., Hill, R.T., Place, A.R., and Hamann, M.T. The expanding role of marine microbes in pharmaceutical development. *Curr. Opin. Biotechnol.*, 21(6) (2010): 780–786.

Webster, N. and Hill, R. Vulnerability of marine microbes on the Great Barrier Reef to climate change. In: *Climate Change on the Great Barrier Reef*. Johnson, J.E. and Marshall, P.A., Eds. Great Barrier Reef Marine Park Authority and Australian Greenhouse Office, Townsville, Queensland, Australia, 2007, pp. 96–120.

Zhu, W., Chiu, L.C.M., Ooi, V.E.C., Chan, P.K.S., and Ang, P.O. Antiviral property and mode of action of a sulphated polysaccharide from *Sargassum patens* against *Herpes simplex* virus type 2. *Int. J. Antimicrob. Agents*, 24(3) (2004): 279–283.

CHAPTER **6**

Microalgae as a Source of Pigments
Extraction and Purification Methods

*Helena M. Amaro, Isabel Sousa-Pinto,
F. Xavier Malcata, and A. Catarina Guedes*

CONTENTS

6.1 Introduction   99
6.2 Microalgal Pigments   100
    6.2.1 Chlorophylls   101
    6.2.2 Carotenoids   101
    6.2.3 Phycobiliproteins   102
6.3 New Cell Disruption Methods   103
    6.3.1 Pulsed Electric Field for Pigment Extraction   103
    6.3.2 Microwave-Assisted Pigment Extraction   104
6.4 Microalgal Pigment Extraction and Purification Methods   104
    6.4.1 Chlorophyll Extraction   104
        6.4.1.1 Classical Solvent Extraction of Chlorophylls   105
        6.4.1.2 Supercritical Fluid Extraction of Chlorophylls   107
        6.4.1.3 Purification and Chlorophyll Quantification Methods   108
    6.4.2 Carotenoid Extraction   108
        6.4.2.1 Classical Solvent Extraction of Carotenoids   108
        6.4.2.2 Pressurized Fluid Extraction of Carotenoids   113
        6.4.2.3 Supercritical Fluid Extraction of Carotenoids   115
    6.4.3 Classical Solvent Extraction and Purification of Phycobiliproteins   118
Acknowledgments   122
References   122

6.1 INTRODUCTION

The marine environment constitutes two-thirds of our planet and holds the major part of biodiversity. The extensive resources provided thereby constitute the basis of many economic activities, and the marine environment offers a wide array of applications for biotechnology in the coming future (Guedes et al., 2011b). One nuclear element of marine ecosystems is microalgae. They constitute a diverse group of microscopic prokaryotic and eukaryotic photosynthetic organisms with vital ecological importance; some of the most efficient converters of solar energy to valuable products, at the expense of (theoretically) inexpensive natural resources, are indeed microalgae. Microalgae (especially those from marine origin) remain to date largely unexplored. They can be classified according to their colors due to the prevalence of certain pigments—Chlorophyceae (green color), Rhodophyceae (red color), Cyanophyceae (blue green), and Pheophyceae (brown). The major microalgal pigments include chlorophyll *a*, *b*, and *c*, β-carotene, phycocyanin (PC), xanthophylls, and phycoerythrin (PE) (Dufossé et al., 2005). Microalgae have thus attracted commercial interest due to their potential to generate valuable products, namely, pigments; some of them have met wide success, for example, lutein, astaxanthin, β-carotene, phycobiliproteins, and PCs (Porra, 1991; Guedes et al., 2011a,b; Amaro et al., 2013; Cuellar-Bermudez et al., 2015).

One of the major bottlenecks in obtaining molecules from microalgae is the difficulty of extracting some metabolites, which can compromise results of high-throughput screening analysis. The development of extraction techniques for microalgae has become a field of growing interest in the scientific community. Most techniques that have been developed are intended for heat-stable molecules and thus seldom suitable for high-throughput screening of sensitive molecules. Concerning the latter, a few studies have been carried out to extract natural bioactive products from microorganisms since the beginning of the 1980s. Many extraction techniques have accordingly been developed (Serive et al., 2012). Pigment studies, in particular, have been steadily increasing since the early 1970s, both in the field of oceanography (Jeffrey et al., 1997; Szymczak-Żyła et al., 2011) and for industrial applications (Hejazi et al., 2002; Machmudah et al., 2006). These techniques are often suitable for molecule purification from large amounts of biomass (Hosikian et al., 2010), but not for high-throughput screening purposes.

Microalgae species are characterized by a huge biodiversity. This biodiversity includes thick-walled green or red algae, silicified diatoms, cyanobacteria with multilayered walls, red algae with wall-bound exopolysaccharides, and armored dinoflagellates, which need to be broken before extraction is possible (Porra, 1991). Another important requirement is easy solubilization of molecules of a wide polarity range. At laboratory scale, it is tempting to use strong solvents to extract targeted molecules. However, acetone, chloroform, dimethylacetamide, dimethylformamide (DMF), dimethyl sulfoxide (DMSO), and methanol are unsuitable at industrial scale due to safety considerations (low lethal dose, carcinogenic, harmful, irritant, or toxic features) (Jeffrey et al., 1997). Eco-friendly approaches can be used instead, via mild solvents that would limit health risks while facilitating scale-up. The extraction method needs to be quick and simple to use and require no heavy equipment.

The main pigments produced by microalgae will be briefly discussed in this chapter, in terms of relevance for industry, main extraction and coextraction techniques employed, and associated methods of purification.

6.2 MICROALGAL PIGMENTS

As mentioned before, microalgae constitute a heterogeneous group of microorganisms. Among the different phyla, cyanobacteria are oxygenic photosynthetic prokaryotes showing large diversity in morphology, physiology, ecology, biochemistry, and other characteristics. Different phyla of microalgae contain different pigments (see Table 6.1).

Chlorophylls, carotenoids (carotenes and xanthophylls), and phycobilins are the three major classes of photosynthetic pigments found in microalgae. Chlorophylls and carotenoids are generally fat-soluble molecules, whereas phycobilins are water-soluble. For a better understanding, the main features of each one are presented.

TABLE 6.1   Main Pigments Found in the Different Phyla of Microalgae

Phylum	Common Name	Pigments	References
Chlorophyta	Green microalgae	Chlorophyll *a*, *b*, β-carotene, prasinoxanthin, siphonaxanthin, astaxanthin	Van den Hoek et al. (1995), Lee (1999), Graham and Wilcox (2000)
Diatomophyceae/ diatoms	Brown microalgae	Chlorophyll *a* and *c*, β-carotene, fucoxanthin, diadinoxanthin	Begum et al. (2015)
Cryptophytes	Cryptomonads	Chlorophyll *a* and *c*, carotenoids, phycobiliproteins	Van den Hoek et al. (1995), Lee (1999), Graham and Wilcox (2000)
Cyanobacteria	Blue-green microalgae	Chlorophyll *a*, xanthophylls, phycobiliproteins	Graham and Wilcox (2000)
Euglenophyta	Euglenoids	Chlorophyll *a* and *b*, diadinoxanthin, neoxanthin, β-carotene	Van den Hoek et al. (1995), Lee (1999), Graham and Wilcox (2000)
Dinophyta	Dinoflagellates	Chlorophyll *a* and *c*, β-carotene, peridinin	Van den Hoek et al. (1995), Lee (1999), Graham and Wilcox (2000)

Source: Adapted from Begum, H. et al., *Crit. Rev. Food Sci. Nutr.*, 2015.

6.2.1 Chlorophylls

Chlorophyll is the main pigment in microalgae and is responsible for light harvesting—an essential piece of photosynthesis. Its molecules are arranged in and around the photosystems in the thylakoid membranes of chloroplasts (Hosikian et al., 2010). There are two main types of chlorophyll, chlorophyll *a* and chlorophyll *b*; the main difference is a methyl group in chlorophyll *a* replaced by a formyl group in the latter (Scheer et al., 2004).

Chlorophyll is one of the valuable bioactive compounds that can be extracted from microalgal biomass. Chlorophyll *a* is used as a natural food coloring agent due to its stability. Microalgae, namely, the cyanobacterium *Spirulina platensis*, hold the advantage of only producing chlorophyll *a*. Therefore, the *Spirulina* sp. is the largest source of chlorophyll suitable for colorants, in cosmetic and pharmaceutical products (Begum et al., 2015). Additionally, this molecule exhibits good antioxidant capacity, as well as antimutagenic properties—with proven effects in chronic ulcer treatment and in the wound healing process (Hosikian et al., 2010).

6.2.2 Carotenoids

There are more than 400 carotenoids in nature, but β-carotene, lutein, and astaxanthin are the most known and bearing the highest relevance in terms of health benefits and industrial applications (Begum et al., 2015). Carotenoids are terpenoid pigments derived from a 40-carbon polyene chain, providing a distinctive molecular structure in terms of light absorption (Del Campo et al., 2007). Carotenoids may be complemented by cyclic groups and oxygen-containing functional groups. Hydrocarbon carotenoids are named as carotenes as a whole, whereas oxygenated derivatives are known as xanthophylls—with oxygen appearing as a hydroxyl group (e.g., lutein), oxi-group (e.g., canthaxanthin), or a combination of both (e.g., astaxanthin) (Del Campo et al., 2007). There are two types of carotenoids, namely, primary and secondary carotenoids. Structural and functional components of the cellular photosynthetic apparatus are the primary ones (i.e., xanthophylls), whereas secondary carotenoids include those produced by microalgae to large levels after exposure to specific environmental stimuli (Eonseon et al., 2003). Xanthophylls are relatively hydrophobic molecules, so they are typically linked to membranes or involved in noncovalent binding to specific proteins—but usually localized in the thylakoid membrane. Secondary carotenoids are found in lipid vesicles (Grossman et al., 1995). Carotenoids can be extracted via such organic solvents as acetone, methanol, or DMSO (Begum et al., 2015).

β-Carotene exerts numerous benefits with the human body; it helps in immunity and cataract prevention, night blindness, skin diseases, and cancer prevention, in addition to its use as food colorant (Begum et al., 2015).

Astaxanthin has provitamin activity, as is the case of β-carotene. Its antioxidant activity has been reported under both hydrophilic and hydrophobic conditions. (Jyonouchi et al., 1994; Begum et al., 2015). Astaxanthin also proved effective in the prevention of arteriosclerosis, coronary heart disease, ischemic brain development, and neuronal damage, besides immunomodulatory effects (Guedes et al., 2011a; Begum et al., 2015). Astaxanthin can be used in animal feed for flesh pigmentation of Atlantic salmon and rainbow trout, besides enhancing the immunity of fish and shrimp toward efficient growth and survival thereof (Guedes and Malcata, 2012).

Lutein is a xanthophyllic compound located in the microalgal chloroplast. It is used as food colorant by EU (E-161 b) and also plays a role in preventing retinal degeneration, some types of cancer, and cardiovascular diseases due to its antioxidant capabilities (Guedes et al., 2011a).

6.2.3 Phycobiliproteins

Phycobiliproteins are light-harvesting pigments commonly present in Cyanophyceae and Cryptophyceae (Glazer, 1994). They are assembled in phycobilisomes and are attached to the surface of thylakoids for photosynthesis. According to Bryant et al. (1979), there are four main classes of phycobiliproteins, namely, allophycocyanin (APC, bluish-green), PC (blue), PE (red), and phycoerythrocyanin (PEC, orange), in cyanobacteria and red microalgae. In most cyanobacteria, the main phycobiliprotein is C-phycocyanin (CPC). This protein can be isolated from wild-type cells of the blue-green microalgae *Oscillatoria agardhii* (Glazer, 1994; Begum et al., 2015). Phycobiliproteins are used at commercial scale, as natural dyes and food colorant; this is the case of phycocyanin from *S. platensis* or PE from *Phorphyridium aerugineum* in cosmetic items (Santiago-Santos et al., 2004). Moreover, PC is used as a pharmaceutical agent because of its antioxidant, anti-inflammatory, neuroprotective, and hepatoprotective properties (Sekar and Chandramohan, 2008), also observed in that obtained from *Aphanizomenon flos-aquae* (Benedetti et al., 2004). In addition, CPC derived from *S. platensis* appears to possess a strong hypocholesterolemic activity and an anticancerous effect against human leukemia K562 cells (Liu et al., 2000; Nagaoka et al., 2005). APC also demonstrated to inhibit enterovirus 71–induced cytopathic effects (Begum et al., 2015).

6.3 NEW CELL DISRUPTION METHODS

Bioproducts produced by microalgae are often located intracellularly and accumulated in organelles (e.g., pigments) and vesicles, or else in the cytoplasm. The presence of a cell wall surrounding the cells, and especially of an intact cytoplasmic membrane that acts as a semipermeable barrier, influences extraction of these compounds from cells (Vanthoor-Koopmans et al., 2013). Traditionally, extraction of microalgal bioproducts has been mainly performed from dried biomass with organic or aqueous solvents, depending on the polarity of the compound at stake (Cerón et al., 2008). Conventional liquid extraction of compounds from microalgal matrices is time-consuming, and a relatively large amount of solvents has to be used; in the case of organic solvents, this is expensive and potentially harmful. In order to reduce processing time and solvent volumes, cells are mechanically disrupted prior to extraction. Mechanical disruption of microalgae can be accomplished in a variety of ways, such as bead milling, homogenization, and ultrasound (Prabakaran and Ravindran, 2011). However, these mechanical cell disruption methods are characterized by lack of specificity that causes a range of cell debris and other impurities to be released with the compound of interest. This negatively affects purification downstream (Balasundaram et al., 2009; Luengo et al., 2014). Therefore, new "clean" technologies have been developed, like pulsed electric field (PEF) and microwave-assisted extraction (MAE), which aid in the purification steps.

6.3.1 Pulsed Electric Field for Pigment Extraction

Treatment of fresh microalgal biomass by PEF or electroporation may be an alternative to conventional cell disruption techniques. PEF technology causes cell membrane permeabilization when cells are exposed to an intense electric field of short duration (milli- to microseconds). It undergoes an electrical breakdown that increment cell membrane permeability to ions and macromolecules due to formation of local defects or pores in the cell membranes—thus rendering them permeable to molecules that would otherwise be unable to cross it. Depending on the processing parameters selected, the membrane can either become transiently or permanently permeable, making electroporation either reversible or irreversible. In reversible electroporation, pores created by the electric field are able to reseal after treatment, whereas in irreversible electroporation the pores in the cytoplasmic membrane remain as permanent. Application of PEF to improve extraction of compounds of interest from microalgae requires irreversible, rather than reversible electroporation (Luengo et al., 2014).

Several studies have demonstrated the potential of PEF to enhance extraction of compounds, such as lipids and carotenoids, from fresh microalgal biomass (Goettel et al., 2013; Zbinden et al., 2013; Grimi et al., 2014; Luengo et al., 2014). There is evidence that application of electroporation to cells of *Chlorella vulgaris* under a 20 kV cm^{-1} field for 75 µs significantly increased extraction yield for carotenoids, and chlorophylls *a* and *b* increased 1.2, 1.6, and 2.1 times, respectively (Luengo et al., 2014). Furthermore, it has been proven that increasing temperature increases the sensitivity of *C. vulgaris* cells to irreversible electroporation. PEF treatments applied at temperatures above 30°C, at 25 kV cm^{-1} for 100 µs and 25°C–30°C, were able to increase lutein extraction yield ca. 3.5–4.2-fold compared to the control—thus resulting in the most suitable treatment conditions for maximizing lutein extraction at lowest energy cost (Luengo et al., 2015). An increase of pigments (chlorophylls and carotenoids) can be also achieved with organic solvents due to their better solubility in organic chemicals (Devesa et al., 2007). A proposed two-stage PEF-assisted procedure, proposed by Parniakov et al. (2015), also allows effective extraction from the *Nannochloropsis* spp. using less concentrated mixtures of organic solvents with water.

6.3.2 Microwave-Assisted Pigment Extraction

In the last 10 years, there has been an increased interest in extraction techniques involving microwaves. MAE and vacuum microwave-assisted extraction (VMAE) have been proposed as efficient and rapid processes to extract antioxidants or pigments from plants, thus allowing reduced solvent consumption and shorter extraction times, with equivalent or higher extraction yields (Pasquet et al., 2011). Choi et al. (2007) reported the MAE of astaxanthin from the red yeast *Xanthophyllomyces dendrorhous* and showed that combining microwave irradiation to destroy cell walls with solvent extraction allows a better pigment recovery. Pasquet et al. (2011) compared the performance of microwave irradiation (MAE and VMAE) to extract pigments from two marine microalgae, *Dunaliella tertiolecta* and *Cylindrotheca closterium*, with conventional processes (cold and hot soaking and ultrasound-assisted extraction). MAE was found to be the best extraction process for *C. closterium* pigments—as it combines rapidity, reproducibility, homogeneous heating, and high extraction yields.

6.4 MICROALGAL PIGMENT EXTRACTION AND PURIFICATION METHODS

Pigment extraction processes applied to microalgae are mainly derived from phytochemical techniques developed for higher plants and macroalgae. The main parameters driving the selection of an extraction technology are biochemical characteristics of extracted molecules, velocity, amount of solvent use, reproducibility, extraction yield, selectivity, protection of extracted molecules against chemical transformation, dimension, cost, and easiness (Wang and Weller, 2006; Pasquet et al., 2011).

As emphasized before, there are other parameters that affect the efficiency of organic solvent extraction further to cell disruption—including storage conditions of microalgae biomass, organic solvents used, extraction time, and number of extraction steps (Pasquet et al., 2011).

The most usual methods employed in the process of extraction and purification of microalgal pigments will be presented in the following.

6.4.1 Chlorophyll Extraction

Usually, chlorophyll is extracted from dried biomass. However, in the case of marine microalgae, the process of extracting chlorophyll begins with dewatering and desalting the highly dilute microalgal culture (biomass concentration of 0.1%–1% w/v). Then, extraction by organic solvent or supercritical fluid will ensue (Hosikian et al., 2010). It should be taken into account that the amount of chlorophyll extracted from a particular microalgal species is highly dependent on its growth stage; extraction in the stationary growth phase has been shown to lead to significantly higher amount of chlorophyll compared to the same microalgal species in the logarithmic phase (Hosikian et al., 2010).

6.4.1.1 Classical Solvent Extraction of Chlorophylls

The most traditional method employed is organic solvent extraction (Jeffrey et al., 1997; Simon and Helliwell, 1998). The extraction process involves organic solvent penetration through the cell membrane and dissolution of lipids as well as lipoproteins from chloroplast membranes (Jeffrey et al., 1997). As stated before, previous cell disruption, achieved through grinding, homogenization, ultrasound, or sonication, significantly improves the effectiveness of chlorophyll extraction using organic solvents (Jeffrey et al., 1997; Simon and Helliwell, 1998; Macías-Sánchez et al., 2009a; Hosikian et al., 2010). Chlorophyll is highly reactive; the yield of a particular extraction procedure is also affected by the formation of degradation products when their molecules are exposed to excess light, oxygen/air, high temperatures, and acidic or basic conditions (Jeffrey et al., 1997; Cubas et al., 2008).

A comparison of various studies on organic solvent extraction of microalgal chlorophyll is depicted in Table 6.2. Sartory and Grobbelaar (1984), Jeffrey et al. (1997), and Simon and Helliwell (1998) found methanol and ethanol to be extraction solvents superior to acetone. Simon and Helliwell (1998) conducted their sonication-assisted chlorophyll extractions in an ice bath and in the dark to prevent formation of degradation products. They found that, with sonication, methanol removed three times more pigment than 90% acetone. Additionally, when tissue grinding was used, methanol removed 20% more pigment than 90% acetone. Sartory and Grobbelaar (1984) similarly found that 90% acetone was an inefficient organic solvent compared to methanol or 95% ethanol. However, it has been shown that the use of methanol induces formation of chlorophyll degradation products (Mantoura and Llewellyn, 1983; Jeffrey et al., 1997). Although 100% acetone was found not to yield the highest amount of chlorophyll from any particular species, its use strongly inhibited the decay of that pigment (Jeffrey et al., 1997). In studies carried out by Jeffrey et al. (1997) and Macías-Sánchez et al. (2009a), DMF was found to be an extraction solvent better than methanol, 90% ethanol, 100% ethanol, or 90% acetone. Extraction with DMF does not require cell disruption, and pigments were completely extracted after a few steps of soaking; the pigments remained stable for up to 20 days, when stored in the dark at 5°C. However, DMF's toxic nature decreased its appeal as an efficient solvent (Jeffrey et al., 1997; Schuman et al., 2005). It was also found that storage of the microalgae at low temperatures after filtration (−18°C or −20°C) assisted cell disruption and promoted extraction of chlorophyll; this is also the case of freezing the biomass in liquid nitrogen, followed by lyophilization and then storage at −18°C (Jeffrey et al., 1997; Schumann et al., 2005).

Sartory and Grobbelaar (1984) found the efficiency of chlorophyll extraction from freshwater microalgae to be optimal when the filtered microalgae and solution were refluxed at the solvent's boiling point. It was shown that boiling for 3–5 min in methanol or 95% acetone prior to 24-h extraction led to quantitative isolation of chlorophyll without formation of degradation products. However, when the mixture was subjected to 100°C, degradation products started to form (Sartory and Grobbelaar, 1984). Such findings are contradictory with the general assumption that chlorophyll degrades upon slight temperature elevation.

TABLE 6.2   Comparison of Classic Organic Solvent Extraction of Microalgal Chlorophyll

Microalga Species	Solvents Tested	Cell Disruption	Main Conclusions	Reference
Phytoplankton	Methanol (90%), ethanol (90%), ethanol (100%), DMF	—	DMF is superior to all other solvents and cell disruption improves extraction in all cases.	Jeffrey (1968)
Dunaliella salina	DMF, methanol	Ultrasound	DMF was found more efficient than methanol.	Macías-Sánchez et al. (2009)
Scenedesmus quadricauda, Selenastrum capricornutum, Microcystis aeruginosa	Ethanol (95%), methanol (95%), acetone (90%)	Homogenization, sonication, soiling	1. Methanol and 95% ethanol were superior to 90% acetone. 2. Boiling the algae in either methanol or 95% ethanol for 5 min and allowing extraction for 24 h resulted in complete extraction of pigments without formation of degradation products.	Sartory and Grobbelaar (1984)
Stichococcus, Chlorella	Acetone and DMF	Grinding, ultrasound, bead beater	1. DMF was found to be the most efficient solvent. 2. Acetone extracted 56%–100% of chlorophyll *a* extracted by DMF. 3. DMF does not require cell disruption. 4. Freeze-drying before analysis aids in extraction.	Schuman et al. (2005)
Freshwater algae, *Selenastrum obliquus*	Methanol and acetone	Probe sonication, bath sonication, tissue grinding, mortar and pestle	1. Under sonication, methanol removed three times more pigment than acetone. 2. After tissue grinding, methanol removed 20% more than acetone.	Simon and Helliwell (1998)

6.4.1.2 Supercritical Fluid Extraction of Chlorophylls

Supercritical fluid extraction (SFE) started to be thoroughly investigated in the 1960s (Herrero et al., 2010). SFE has many advantages over organic classical solvent extraction, standing out the high purity of the extract. In addition to requiring less processing steps, SFE is significantly safer than organic solvent extraction and can be carried out at moderate temperatures—thus minimizing extract degradation and, in the absence of light, beyond the many other advantages shown in Table 6.3 (Sahena et al., 2009).

The supercritical state is achieved when a substance is exposed to conditions exceeding its critical temperature (T_c) and pressure (T_p). In this state, the substance has liquid-like densities with gas-like viscosities. The solvent power of supercritical fluids is the highest for slightly polar or nonpolar components but is lower for analytes with higher molecular weights. The fluid is easily removed from the extract through expansion to atmospheric pressure (Brunner, 2005). CO_2 is the most commonly used fluid for SFE, as it is inexpensive, nonflammable, readily available, and somewhat inert. The moderate nature of CO_2 yields higher quality extracts, because it avoids excessive heating that often leads to degradation, and thus allowing extraction of thermolabile compounds (Mendes et al., 2003; Macías-Sánchez, 2005; Guedes et al., 2013).

Supercritical fluid extraction with CO_2 ($SCCO_2$) is also preferred due to its high diffusivity and the ease in manipulating its solvent strength. The addition of a polar solvent, like methanol or water, to supercritical CO_2 allows it to extract polar compounds (Mendes et al., 2003; Macías-Sánchez et al., 2005).

Chlorophyll is mainly used in food technology, so there are stringent regulations regarding its quality; consequently, using $SCCO_2$ extraction permits one to obtain a solvent-free, highly pure extract. The chlorophyll extraction in $SCCO_2$ from microalgae depends on the fluid density, which is in turn a function of operating pressure and temperature. Studies have been conducted on $SCCO_2$ chlorophyll extraction from at least two microalgal species, *Nannochloropsis gaditana* and *Synechococcus* sp. Optimum extraction conditions were found to be 60°C and 400 bar for *N. gaditana* and 60°C and 500 bar for *Synechococcus* sp. (Hosikian et al., 2010).

TABLE 6.3 Advantages in Use of Supercritical Fluid Extraction over Classical Organic Solvent Extraction

Organic Solvent Extraction	SFE-CO_2 Extraction
Time-consuming, high expenditure of solvents	Quick, few nonsolvent requirements, environment friendly
High expenditure of solvents; subsequent extraction solvent removal is required thus increasing operational cost	$SCCO_2$ extraction produces solvent-free extracts
Heavy metal and inorganic salt may be coextracted from the raw material or be present in the solvents	$SCCO_2$ extract is totally free of heavy metals and inorganic salts, since they are neither extractable nor present in the CO_2 extractant or the SFE system; nontoxic and nonflammable
Polar substances may be dissolved in lipophilic substances from the raw material, due to the low selectivity of the solvent	Due to its nonpolar nature, $SCCO_2$ does not extract polar substances unless a polar cosolvent is added

6.4.1.3 Purification and Chlorophyll Quantification Methods

Chromatographic techniques are used to fractionate/isolate/quantify chlorophylls. The three types of chromatography that have been more widely used are paper chromatography, thin-layer chromatography (TLC), and high-pressure liquid chromatography (HPLC). Paper chromatography was mostly used in the 1950s and 1960s, during the early stages of the development of chromatographic techniques. TLC emerged as an easy quantitatively recovering technique, using silica gel as stationary phase and allowing complete separation of pigments; however, it permits formation of degradation products (Jeffrey, 1968).

Later, HPLC techniques have risen with superior performance to separate, identify, and quantify compounds. It is faster, often coupled with automatic detection systems, more precise, and with higher degree of sensitivity (Shoaf, 1978). Reverse-phase HPLC is preferred to normal phase because it does not separate polar compounds, and its polar stationary phase promotes pigment degradation (Mantoura and Llewellyn, 1983). Regular-phase HPLC possesses a relevant drawback: it is not compatible with aqueous samples, whereas many of the solvents used for chlorophyll extraction from microalgae are aqueous-based (Mantoura and Llewellyn, 1983). Several HPLC configurations have been employed, with each providing separation of pigments to varying extents and distinct resolutions (Jeffrey et al., 1997). There are different types of detectors that may be used to measure the concentrations of separated pigments. The most commonly used rely on fluorescence and absorbance. Jeffrey et al. (1997) stated that fluorescence detection is more sensitive and more selective than absorbance when used to analyze chlorophylls among carotenoids. Table 6.4 summarizes previous studies on HPLC separation and purification of microalgal pigments.

6.4.2 Carotenoid Extraction

As happens with chlorophyll, carotenoids are extracted from dried microalgal biomass. The green algae *Haematococcus pluvialis* is one of the most important biological source of astaxanthin. Other microalgae species that are able to accumulate secondary carotenoids are *Botryococcus braunii*, *Chlamydomonas nivalis*, *Chlorella* sp., *Chlorococcum* sp., *Chloromonas nivalis*, *Coelastrella striolata* var. *multistriata*, *Dunaliella* sp., *Eremosphaera viridis*, *Euglena* sp., *Neochloris wimmeri*, *Scenedesmus* sp., *S. obliquus*, *S. komarekii*, *Scotiellopsis oocystiformis*, *Protosiphon botryoides*, *Tetracystis intermedium*, and *Trachelomonas volvocina* (Cuellar-Bermudez, 2015). In contrast to the said species, astaxanthin content in *H. pluvialis* represents 90% of the total carotenoids. Therefore, extraction of this compound from other species represents a disadvantage in many markets due to the purification steps required and associated processing costs (Borowitzka, 2013). The methods most widely applied for carotenoid extraction use classical solvent extraction, pressurized fluid extraction (PFE), and SFE.

6.4.2.1 Classical Solvent Extraction of Carotenoids

In classical solvent extraction, multiple different methods are employed depending on the carotenoid—as apparent in Table 6.5. Astaxanthin classic extraction methods may include use of organic solvents, breakdown pretreatment process of encysted cells (cryogenic grinding and acid/base treatment), enzyme lysis (kitalase, cellulose, and abalone acetone powder, mainly β-glucuronidase), mechanical disruption, and spray drying (Sarada et al., 2002; Kang and Sim, 2007). Astaxanthin accumulated in *H. pluvialis* is

TABLE 6.4 Chromatographic Separation of Microalgal Pigments

Method	Stationary Phase	Mobile Phase	Pigment Separation	Reference
HPLC	3 μm C18 Pecosphere	90:10 (v/v) methanol/acetone for 8 min Flow rate: 1 mL min^{-1} Preinjection mix of sample 3:1 (v/v) sample/0.5 M ammonium acetate	Isocratic separation of chlorophyll a from other pigments and compounds	Jeffrey et al. (1997)

		Solvent A: 80:20 (v/v) methanol/0.5 M ammonium acetate
		Solvent B: 90:10 methanol:acetone
		Elution order:
		  0–3 min: solvent A
		  3–17 min: solvent B
		Flow rate: 1 mL min^{-1}
		Preinjection mix of sample 3:1 (v/v) sample: 0.5 M ammonium acetate
			Isocratic separation of chlorophylls a, b, and c and other derivative products	Wright and Shearer (1984)

		Solvent A/80:20 (v/v) methanol: 0.5 M ammonium acetate
		Solvent B: 90:10 (v/v) acetonitrile/water
		Solvent C: ethyl acetate
		Elution order:
		  0–4 min: linear gradient from 100% A to 100% B
		  4–18 min: linear gradient to 20% B and 80% C
		  18–21 min: linear gradient to 100% B
		  21–24 min: linear gradient to 100% A
		  24–29 min: isocratic flow of 100% A
			Ternary gradient able to extract over 50 pigments	

	C18 Octadecyl silica	Linear gradient from 90% acetonitrile to ethyl acetate for 20 min at a flow rate of 2 mL min^{-1}	High-resolution separation of carotenes, chlorophylls, xanthophylls, and their degradation products	

(Continued)

TABLE 6.4 (*Continued*) Chromatographic Separation of Microalgal Pigments

Method	Stationary Phase	Mobile Phase	Pigment Separation	Reference
	C3 Zorbax, C8 Zorbax, C18 Zorbax, Shandon Hypersil ODS	Solvent P (ion pairing reagent): 1.5 g tetrabutyl ammonium acetate and 7.7 g of ammonium acetate, dissolved in 100 mL of water. Solvent A: solvent P/water/methanol (10:10:80 v/v/v) Solvent B: 20:80 (v/v) acetone/methanol Elution order: 0–10 min: linear gradient from A to B 10–22 min: isocratic flow of B	High-resolution separation of chlorophylls and all major pigments. High recovery (over 90%) of pigments	Mantoura and Llewellyn (1983)
HPLC	Sep-Pak C18 Bondapak	Solvent A: 97% methanol Solvent B: 97% acetone Elution order: 0–15 min: 100% solvent A 15–20 min: linear gradient to 77% solvent A/23% solvent B Isocratic flow of 77% solvent A/23% solvent B Flow rate: 1 mL min^{-1}	High-resolution separation of chlorophylls and derivatives Recovery of total pigments by HPLC greater than 96%	Sartory (1985)
	Glucose	30 mL of 1:1 (v/v) diethyl ether/petroleum spirit	Spectrophotometric analysis found to overestimate chlorophyll *c* by up to 22% and to underestimate chlorophyll *b* by 10%–20% Chlorophyll *a* correctly quantified	Madgwick (1996)

(*Continued*)

TABLE 6.4 (*Continued*) Chromatographic Separation of Microalgal Pigments

Method	Stationary Phase	Mobile Phase	Pigment Separation	Reference
TLC	Silica gel	*Solvent system 1* (modified Bauer solvents): first dimension is benzene/petroleum ether/acetone (10:2.5:2 v/v/v). Second dimension is benzene/petroleum ether/acetone/methanol (10:2.5:1:0.25 v/v/v) *Solvent system 2*: First dimension is benzene/petroleum ether/acetone/methanol (10:2.5:1:0.25 v/v/v). Second dimension is petroleum ether/acetone/n-propanol (8:2:0.5 v/v/v) *Solvent system 3*: First dimension is benzene/petroleum ether/acetone (10:2.5:2 v/v/v). Second dimension is petroleum ether/acetone/n-propanol (8:2:0.5 v/v/v)	Eight major pigments, as well as 8–10 minor derivatives successfully separated	Lynn Co and Schanderl (1967)
		Petroleum ether/ethyl acetate/diethyl amine (58:30:12 v/v/v)	All plant pigments separated, except for some minor components	Riley and Wilson (1965)
	Sucrose	First dimension: 0.8% *n*-propanol in light petroleum (by volume) Second dimension: 20% chloroform in light petroleum (by volume)	Complete separation of chlorophylls and carotenoids	Jeffrey (1968)

TABLE 6.5   Microalgal Carotenoid Classical Solvent Extraction

Carotenoids	Microalgae Source	Extraction/Purification Method	Yield/Extraction Efficiency	Reference
Astaxanthin	*Chlorococcum* sp.	Methanol/dichloromethane 1:3 (v/v). Cells disrupted by French pressure at 110 MPa	Yield: 7.09 mg g_{DW}^{-1}	Ma and Chen (2001)
		Saponification in darkness (50 mg NaOH in 100 mL methanol)		
	Haematococcus pluvialis	Cell acid digestion with HCl 2 M. Acetone extraction at 70°C for 1 h	Efficiency: 87%	Sarada et al. (2002)
		Dodecane mixing for 48 h. Saponification with methanolic NaOH (0.02 M). Sedimentation in darkness at 4°C, 12 h	Efficiency: 85%	Kang and Sim (2007)
		Hexane/acetone/ethyl alcohol (100:70:70% v/v) extraction	N/A	Domínguez-Bocanegra et al. (2004)
		DMSO extraction at 55°C, vortex 30 s	N/A	Orosa et al. (2005)
		Dodecane/methanol (1:1) (v/v), 2-step procedure, and addition of NaOH at 25°C	Efficiency: 85%	Mäki-Arvela et al. (2014)
β-Carotene	*Dunaliella salina*	Hexane/acetone/EtOH (2:1:1) (v/v) at 25°C for 24 h	Efficiency: 90.42%	Hu et al. (2008)
		Followed by KOH saponification		
Carotenoids	*D. salina*	DMF extraction at 25°C for 3 min of sonificationStorage at 4°C	Yield: 27.7 mg g^{-1}	Macias-Sanchez et al. (2009)
Lutein	*Scenedesmus obliquus*	Bead beater pretreatment, extraction with diethyl ether at 25°C in S/R ratio, 2, 5 extraction steps	Efficiency: 99% Yield: 2.05 mg g^{-1}	Chan et al. (2013)
Fucoxanthin	*Phaeodactylum tricornutum*	Freeze-dried cells, extraction with ethanol during 60 min	Yield: 15.33%	Sang et al. (2012)

ca. 95% in esterified form, and it usually requires hydrolysis step to free the astaxanthin molecules. Sarada et al. (2002) tested the extractability of carotenoids from *H. pluvialis* with hydrochloric acid (2 N) for 10 min at 70°C, followed by acetone extraction for 1 h—thus extracting 87% (w/w) of astaxanthin without affecting its composition. Kang and Sim (2007) developed a two-stage solvent procedure with dodecane and methanol to extract free astaxanthin from *H. pluvialis* cells, by mixing the solvents with the culture broth, next to sedimentation of the mixture for 48 h. Later, the dodecane extract was separated from the cell debris, placed in another tank, and mixed with NaOH in methanol (0.02 M) at a volume ratio of 1:1 (to promote saponification of astaxanthin esters to free form). Then, the tank was kept in darkness at 4°C (12 h) to support astaxanthin extraction in the methanol phase. The results indicated a total recovery yield of free astaxanthin over 85% dry weight (DW).

Microwave and ultrasound effects upon the stability of synthetic astaxanthin isomers were studied by Zhao et al. (2006)—who concluded that microwaves induce conversion of other astaxanthin isomers, while ultrasound degrades this pigment into colorless compounds because of cavitation produced in the solvent from propagation of ultrasonic waves.

6.4.2.2 Pressurized Fluid Extraction of Carotenoids

PFE is a method that operates with conventional solvents at controlled temperatures and pressures and has been well established for environmental analysis (Denery et al., 2004). PFE showed higher or equal extraction efficiencies as compared to traditional solvent extraction, while maintaining the integrity of chemical components (Denery et al., 2004). High pressure typically shortens the extraction time and the amount of solvent used. PFE has been actively studied (Plaza et al., 2008; Denery et al., 2004; Jaime et al., 2010; Koo et al., 2012).

Pressurized liquid extraction of algae has been preferred due to specific benefits and is a powerful tool in nutraceutical industry: possibility to avoid excessive heat, oxygen, and light that cause degradation of sensitive compounds (Plaza et al., 2010), lower amount of solvent needed (Jaime et al., 2010), higher selectivity compared with Soxhlet and ultrasound-assisted extraction (Koo et al., 2012), and shorter time needed for extraction (Jaime et al., 2010). Several solvents have been investigated for pressurized liquid extraction of algae, such as ethanol, 2-propanol, hexane, petroleum ether, and water. Ethanol has been one of the best solvents, giving both high yields (Rodríguez-Meizoso et al., 2008; Koo et al., 2012) and extracts with high antioxidative capacity (Herrero et al., 2005)—as can be seen in Table 6.6.

Pressurized extraction of zeaxanthin has been investigated with *Chlorella ellipsoidea* under pressurized conditions with different solvents, such as hexane, ethanol, and 2-propanol, and with *C. vulgaris* using ethanol as solvent (Cha et al., 2010; Koo et al., 2012). It was observed that use of hexane and 2-propanol as solvent requires higher temperatures when compared to ethanol (Koo et al., 2012). The high extraction efficiency of ethanol was also observed in the extraction of carotenoids from *Phormidium* in the temperature range 50°C–100°C at 10.3 MPa, compared with those achieved with water or hexane (Rodríguez-Meizoso et al., 2008). Ethanol was also an efficient solvent for extraction of astaxanthin from *H. pluvialis* at 10.3 MPa and 200°C. When nonpolar hexane was applied, low efficiency was achieved due to the polar nature of zeaxanthin. It has been claimed that zeaxanthin is hardly soluble in hexane and petroleum ether (Mäki-Arvela et al., 2014).

TABLE 6.6   Optimal Conditions for Pressurized Liquid Extraction of Carotenoids from Microalgae

Carotenoid	Microalgae Source	Solvent System	Processing Conditions	Efficiency/ Yield (%)	Reference
Lutein	*Synechocystis* sp.	Ethanol	T (°C): 100 P (MPa): 10.3 T (min): 20	Yield: 2.04 mg g^{-1}	Plaza et al. (2010)
	Chlorella vulgaris	20 mL ethanol for 0.5 g microalga	T (°C): 160 P (MPa): 10.3 T (min): 30	Yield: 3.78 mg g^{-1}	Cha et al. (2010)
	Chlorella	Jet mill treated cell extracted with EtOH, 6% KOH	T (°C): 50 P (MPa): 3 T (min): 30	Efficiency: 1.46	Shibata et al. (2004)
Zeaxanthin	*Chlorella ellipsoidea*	Ethanol	T (°C): 115.4 P (MPa): 10.3 T (min): 23.3	Yield: 4.28 mg g^{-1}	Koo et al. (2012)
	Synechocystis sp.	Ethanol	T (°C): 100 P (MPa): 10.3 T (min): 20	Yield: 1.64 mg g^{-1}	Plaza et al. (2010)

When comparing different solvents toward extraction of carotenoids from *S. platensis* at 10.3 MPa in hexane, petroleum ether, ethanol, and water, the highest yields were achieved in ethanol (19.7 wt%), followed by water (10.12 wt%), hexane (4.3 wt%), and petroleum ether (4.0 wt%) (Herrero et al., 2005). On the other hand, the use of polar solvents favors extracts in terms of antioxidant capacity, as is the case of ethanol. Similarly to the work of Herrero et al. (2005), ethanol gave higher extraction yields compared with hexane during extraction of *H. pluvialis* in the temperature range 50°C–200°C at 10.3 MPa.

As said before, using pressurized liquid extraction is possible to shorten extraction time. Typically, a longer extraction time leads to a higher extraction yield; however, in some cases, for example, extraction of lutein, lower yields were achieved with longer extraction times, since lutein was less thermolabile than astaxanthin (Denery et al., 2004). Short extraction time (just 10 min) gave high extraction yield of fucoxanthin from *Phaeodactylum tricornutum* at 100°C in ethanol, whereas in comparative methods, about the same yields were obtained either at room temperature with ultrasound-assisted extraction or in Soxhlet extraction at 80°C, but only by 30 min (Sang et al., 2012).

The temperature can either be dominant or exhibit a minor effect. However, some combinations of temperature and time can promote side reactions, for example, pheophorbide formation from chlorophyll *a* (Cha et al., 2010; Koo et al., 2012). The most efficient solvent for extraction of zeaxanthin is ethanol with the optimum temperature for extraction of 115.4°C, whereas extraction in hexane gave a lower amount of zeaxanthin. When comparing ultrasound-assisted extraction with pressurized extraction of carotenoids at high temperature, it was stated that one benefit for zeaxanthin extraction in the pressurized system is the lowered liquid viscosity (Koo et al., 2012). One drawback of using higher extraction temperatures, namely, in extraction of lutein and β-carotene from *C. vulgaris*, is the aforementioned formation of pheophorbide from chlorophyll *a*.

This formation increased with increasing time, but the highest pheophorbide formation was observed at 60°C—and decreased with increasing temperature, due to deactivation of chlorophyllase at high temperature. Pheophorbide is a Mg^{2+}-free chlorophyll that may cause dermatitis in human skins and food poisoning above 1.6 mg g^{-1} (Mäki-Arvela et al., 2014). Temperature exhibited only a minor effect on astaxanthin yields during pressurized liquid extraction of *H. pluvialis* performed in acetone at 10.3 MPa, within the temperature range 20°C–100°C for 5 min, using three extraction cycles (Denery et al., 2004). The results revealed that astaxanthin yield was nearly unaffected by temperature, whereas a slight decrease of lutein was observed at higher temperature (Denery et al., 2004). Furthermore, no *trans* to *cis* isomerization of astaxanthin occurred at 40°C and 10.3 MPa. It was finally stated that the pressurized liquid extraction required only half the amount of solvent and 20 min extraction time, when compared with traditional extraction for 90 min.

6.4.2.3 Supercritical Fluid Extraction of Carotenoids

$SCCO_2$ or ethane has been applied to carotenoid separation due to its high selectivity and safety toward thermolabile carotenoids (Krichnavaruk et al., 2008; Jaime et al., 2010). However in some cases, for example, astaxanthin, supercritical CO_2 extraction gives low yields, so use of ethanol as a cosolvent is recommended. However, supercritical extraction of β-carotene with ethane or ethylene has been successfully demonstrated, since β-carotene has higher solubility in these hydrocarbons than in CO_2 (Talisic et al., 2012).

The extraction kinetics for carotenoids entails three different regions: (1) In the beginning, the extraction is linear with time, thus indicating a constant rate, which is caused by either solubility equilibrium or a constant mass transfer resistance; (2) extraction stage rate declines because most of the carotenoids have already been stripped from the solid–liquid interface; and (3) in this region, the extraction rate is very low due to the need for the solvent to diffuse into the algae matrix for residual extraction. The addition of ethanol as a cosolvent can enhance permeabilization and cell rupture and/or swelling of matrix, thus facilitating internal mass transfer (Macias-Sanchez et al., 2009).

The effects of pressure and temperature on supercritical extraction of carotenoids are interrelated; when increasing temperature at relatively low pressure, extraction yield is lowered due to the lower density of CO_2—which also lowers solubility of carotenoids in the solvent, as observed at 70°C and 30 MPa in supercritical CO_2 (Kitada et al., 2009; Bustamante et al., 2011). Otherwise, extraction yields are increased with increasing pressure and temperature, should the carotenoid be sufficiently thermolabile. Solute properties, such as thermal and chemical stability, as well as its polarity and solubility, also affect extraction efficiency (Cardoso et al., 2012).

At high pressures, the carotenoid yield typically increases both with increasing CO_2 pressure and temperature (Bustamante et al., 2011; Aravena et al., 2012). Solubility and vapor pressure of solute are important parameters determining extraction efficiency. Furthermore, viscosity of the solvent decreases with increasing temperature. High temperature enhances the yield of astaxanthin from *H. pluvialis* with supercritical CO_2 when changing temperature from 40°C to 70°C (Aravena and del Valle, 2012)—Table 6.7. This result was explained by the increase in vapor pressure of the solute, thus facilitating mass transfer into the CO_2 phase. Similar results were also achieved by Machmudah et al. (2006). The highest astaxanthin yields from *H. pluvialis* were achieved with pure CO_2 at relatively high temperatures, 60°C–80°C, and pressures about 5 MPa—which is similar to those found by Aravena and del Valle (2012) when starting from dry alga powder.

TABLE 6.7   Optimal Conditions of Supercritical Fluid Extraction for Pigments from Microalgae

Carotenoid	Microalgae Source	Processing Conditions	Achievements	Reference
Astaxanthin	*Haematococcus pluvialis*	EtOH/10% soybean oil extraction at 70°C, 4 MPa	Efficiency: 36%	Krichnavaruk et al. (2008)
		SCCO$_2$/Celite, 25 wt% aqueous homogenate, at 70°C and 5.5 MPa	Efficiency: 60.75%	Aravena et al. (2012)
		SCCO$_2$ at 55 MPa and 343 K	Total extracted, 21.8%; amount extracted, 77.9% AX content in the extract: 12.3%	Machmudah et al. (2006)
		CXE at 7 MPa and 45°C with 50% w/w EtOH content in CO$_2$	Extraction yield: 333.1 mg g$_{DW}^{-1}$ AX content: 62.57 mg g$_{DW}^{-1}$ AX recovery: 124.2% w/w	Reyes et al. (2014)
	Monoraphidium sp.	SCCO$_2$ with acid treatment, EtOH as cosolvent at 20 MPa and 60°C for 1 h	AX yield: 2.45 mg g$_{DW}^{-1}$	Fujii et al. (2012)
Total carotenoids	*Scenedesmus almeriensis*	SFE at 40 MPa and 60°C	$12.17 + 0.24$ µg mg$_{DW}^{-1}$	Macías-Sánchez et al. (2009b)
Lutein	*Sc. almeriensis*	SCCO$_2$ at 30 MPa and 39°C during 300 min	0.0236 mg$_{pigments}$ g$_{DW}^{-1}$	Macías-Sánchez et al. (2010)
	Scenedesmus obliquus	SCCO at 25 MPa, 40°C with a CO$_2$ flow of 2 g min^{-1}	0.028 mg g$_{DW}^{-1}$	Guedes et al. (2013)

AX, astaxanthin; CXE, CO$_2$-expanded ethanol extraction; EtOH, ethanol; MeOH, methanol.

Different optimum temperatures and pressures were obtained for the extraction of β-carotene and zeaxanthin from *Synechococcus* sp. due to the fact that β-carotene is non-polar—whereas zeaxanthin has two hydroxyl groups, different temperature and pressure optima for their extraction have been found (Cardoso et al., 2012). β-carotene yield from *Synechococcus* sp. was the largest at the highest temperature and pressure studied, being

60°C and 40 MPa, respectively, when using pure CO_2 as solvent with ethanol (Cardoso et al., 2012). On the other hand, the highest extraction efficiency for zeaxanthin was achieved with CO_2 at the highest temperature and lowest pressure, 60°C and 20 MPa, respectively (Cardoso et al., 2012). Extraction of β-carotene from *D. salina* has also been demonstrated in supercritical ethane or ethylene (Talisic et al., 2012). Typically, extraction yields increase with increasing density of ethane or ethylene, and ca. 59 wt% yield of β-carotene was achieved within 20 min. On the other hand, the carotenoid yields were quite low at relatively low CO_2 pressure and high temperatures, due to the decrease in density and solubility of carotenoid (Mendes et al., 2003; Kitada et al., 2009; Macias-Sanchez et al., 2009). This effect was apparent in the final yield of lutein using 40 MPa at 70°C; it was lower at 40 MPa than 30 MPa. The solubility of lutein was, however, higher at 40 MPa compared to 30 MPa, thus indicating the presence of diffusion limitations. Furthermore, the initial extraction rates were the same at 30 MPa within 60°C–80°C. This result was explained by the fact that lutein solubility at 80°C was the rate limiting factor, because the density of supercritical CO_2 decreases with increasing temperature at constant pressure (Kitada et al., 2009). Lutein extraction rate and final lutein recovery after 2 h were similarly very low for *C. vulgaris* at 40 MPa and 80°C, whereas at 40°C, lutein recovery was much higher, 0.6% and 1.6%, respectively (Ruen-ngam et al., 2012). In the work by Mendes et al. (2003), a lower carotenoid yield was achieved at 55°C than 40°C, and 20 MPa. The results showed an analogous trend of low solubilities of astaxanthin at three different temperatures and low CO_2 pressure. Analogously, at pressures close to the critical pressure of CO_2, a temperature increase lowers the recovery degree of astaxanthin—since solubility of astaxanthin decreases with increasing temperature, due to a decrease in density (Bustamante et al., 2011). In addition, only a slight increase in astaxanthin extraction efficiency from *H. pluvialis* was observed at 30 MPa of CO_2, when increasing the extraction temperature from 40°C to 60°C.

In some cases, carotenoid yields also decrease with increasing pressure. Supercritical CO_2 has been investigated in carotenoid extraction from *D. salina* (Macias-Sanchez et al., 2009b)—see Table 6.7. For instance, carotenoid yield exhibited a maximum at 40 MPa and 60°C, whereas at a higher pressure, 50 MPa, the yield of carotenoids was much lower. This result was explained by the fact that diffusivity of solvent increases with increasing temperature, while density of CO_2 decreases. Furthermore, vapor pressure of the pigments increased as well (Macias-Sanchez et al., 2009b).

Selectivity for carotenoid extraction is high in algae extraction with pure CO_2, but at the same time, it is very nonpolar, thus limiting the yields of relatively polar carotenoids such as astaxanthin (Mendes et al., 2003; Macias-Sanchez et al., 2009b; Cardos et al., 2012). In some cases, the solubility of *cis* versus *trans* isomers of carotenoid is different in pure CO_2, thus favoring faster extraction of the other isomer (Mendes et al., 2003).

Supercritical extraction of lutein with 30 MPa CO_2 at 60°C resulted in relatively low yield, 0.5 mg g^{-1}, but maximum selectivity, whereas in the presence of ethanol, the yield of lutein was 3 mg g^{-1}—while ca. 9 mg g^{-1} chlorophylls were extracted under the same conditions (Kitada et al., 2009) (see Table 6.7). Supercritical extraction with CO_2 is also very selective in extraction of β-carotene (Cardoso et al., 2012). It was demonstrated that the supercritical extraction of carotenoids gave both carotenoids and chlorophylls, but it was about 18-fold more selective for carotenoid extraction than achieved with ultrasound-assisted extraction with methanol (Macias-Sanchez et al., 2009a).

Solubility difference between *cis* and *trans* isomers of β-carotene in supercritical CO_2 facilitates selective production of the former in supercritical CO_2, since solubility of *cis*-β-carotene is higher than *trans*-β-carotene in CO_2 (Mendes et al., 2003). It is also known that the *cis* form is more easily absorbed by the human body than its *trans* isomer, thus emphasizing the importance of selective recovery of *cis*-β-carotene. When comparing the extraction efficiency of acetone with the efficiency of supercritical CO_2 extraction, a two-fold enhancement of the *cis/trans* ratio was obtained for extraction of β-carotene from *D. salina* using supercritical CO_2 as solvent.

As said before, due to the fact that $SCCO_2$ is a very nonpolar solvent, and xanthophylls (e.g., lutein and astaxanthin) have low solubility, the addition of ethanol aids in the extraction of hydroxyl-containing carotenoids; however, extraction selectivity toward one specific carotenoid is reduced when compared with pure CO_2 (Kitada et al., 2009). Lutein extraction from *Scenedesmus* sp. was found very efficient with CO_2 and ethanol as cosolvent (Yen et al., 2012; Guedes et al., 2013). Lutein yield from *Scenedesmus* sp. increased with increasing CO_2 pressure at 47.5°C, with the maximum recovery being only 3.1% at 40 MPa (Yen et al., 2012). When ethanol was used as entrainer, the yield increased with increasing molar fraction of ethanol up to 62.2%, using 40 mol% ethanol at 40 MPa and 70°C. Therefore, supercritical CO_2 extraction of spray-dried *Scenedesmus* sp. was not feasible without ethanol as cosolvent. Besides, an optimum ethanol concentration in $SCCO_2$ gives the highest carotenoid yield (Yen et al., 2012). The optimum ethanol concentration in the extraction of lutein from *Scenedesmus* sp. was 20 mol%, thus yielding 76.65% recovery at 40 MPa and 70°C, whereas it was only 5% for astaxanthin extraction from *H. pluvialis*—with CO_2 giving the highest astaxanthin yield, of 77.9%, at 70°C and 40 MPa within 240 min (Machmudah et al., 2006; Yen et al., 2012). In the presence of ethanol as cosolvent, the increase of temperature can in some cases have a negative effect—for example, in supercritical CO_2 extraction of astaxanthin from *H. pluvialis* (Bustamante et al., 2011). The reason for these results could have been that the isomerization of astaxanthin in ethanol favors oxidation (Bustamante et al., 2011). Different optimum temperatures and pressures were also found for extraction of β-carotene and zeaxanthin from *Synechococcus* sp. due to their different solubilities in CO_2–ethanol mixture. The optimum temperature for extraction of β-carotene was 40°C at 20 MPa CO_2 and 5 vol% ethanol, whereas for zeaxanthin, the optimum temperature and pressure were 60°C and 20 MPa, respectively (Cardoso et al., 2012).

Another promising method for supercritical CO_2 is to use vegetable oil as a cosolvent. The benefits of this method are higher solubility of, for example, astaxanthin in soybean oil–CO_2 mixture compared with that in pure CO_2 and the possibility to avoid the subsequent separation step of the cosolvent—since the carotenoid can remain in vegetable oil products (Krichnavaruk et al., 2008). This method has been utilized in the preparation of astaxanthin extracted from *H. pluvialis* with supercritical CO_2 using vegetable oils—see Table 6.7 (Krichnavaruk et al., 2008). The optimum amount of soybean cosolvent was 10%, giving 36% extraction efficiency for astaxanthin at 70°C and 40 MPa CO_2.

6.4.3 Classical Solvent Extraction and Purification of Phycobiliproteins

Phycobiliproteins are found in prokaryotic microalgae and eukaryotic red algae. They produce four main classes of phycobiliproteins: APC (bluish-green), PC (blue), PE (purple), and PEC (orange) (Cuellar-Bermudez et al., 2015).

Spirulina sp. is the most studied cyanobacteria in terms of CPC production due to its high protein content—it constitutes up to 20% of its DW. Phycobilisomes from *Spirulina* sp. consist of APC cores, surrounded by CPC peripherally.

PC is a red colored phycobiliprotein, found in the chloroplast of cyanobacteria and red algae. There are two main classes of PE in red microalgae, B-PC and R-PC. Specifically, PE has been extracted and purified from the red algae *Porphyridium cruentum* (Bermejo Román et al., 2002; Benavides and Rito-Palomares, 2004), *Phormidium* sp. A27DM (Parmar et al., 2011), and the thermophile cyanobacterium *Leptolyngbya* sp. KC45. Most studied microalga producers of phycobiliproteins and corresponding extraction and purification steps are presented in Table 6.8. PC must be highly purified in order to meet the standards set by the pharmaceutical or molecular biology fields. Purity is usually determined as the absorbance ratio of A565/A280, which defines the relationship between PC and other contaminating proteins. A purity ratio A565/A280 greater than four corresponds to diagnostics and pharmaceutical grade PC (Benavides and Rito-Palomares, 2004). Some authors use the Abs_{615}/Abs_{565} ratio to determine PC purity relative to PC, which is its closest contaminating protein. Since PC is an intracellular protein, the general purification process relies on three stages: protein extraction by cell disruption, primary recovery, and purification. Disruption methods like sonication, mechanical maceration, and lysozyme treatment have been successfully used to extract PC from microalgae. Choosing the right cell disruption method has a significant impact upon recovery of the overall process.

Most authors carry out a primary recovery step of selective precipitation with ammonium sulfate. This solubility-based method is fast and inexpensive. Bermejo Román et al. (2002) did a single precipitation step with ammonium sulfate at 65% saturation. Parmar et al. (2011) recovered PC from *Phormidium* spp. A27DM via a two-step ammonium sulfate precipitation, at 20% and 70% saturation, while Pumas et al. (2012) treated the cell homogenate of *Leptolyngbya* sp. KC45 with ammonium sulfate at 85% saturation—see Table 6.8.

Purification is a critical step in downstream processing of PC. It is classically achieved by chromatographic methods, such as ion exchange chromatography, hydroxyapatite chromatography, gel filtration, and expanded bed absorption chromatography. However, according to Kawsar and colleagues (2011), hydroxyapatite chromatography is not a reliable method since separation ability depends on the quality of the particles, and regeneration capacity is not good since the material may bind to other contaminating complexes.

As happens with CP, purity ration of APC and CPC can be determined by the A620/A280 ratio. An absorbance ratio ≥0.7 refers to food grade pigment, while reagent and analytical grades correspond to 3.9 and ≥4.0, respectively (Borowitzka, 2013).

The stability of PC after and during extraction is still a bottleneck. Therefore, Chaiklahan et al. (2012) tested the stability of this pigment extracted from *Spirulina* sp., using different temperatures and pH, following addition of certain preservatives (i.e., glucose, sucrose, D-sorbitol, sodium chloride, ascorbic acid, citric acid, sorbic acid, and sodium azide). The best results were obtained at 47°C and pH of 6.0, with a relative concentration (CR) value of 94%. Experiments are run under commercial high-temperature short-time pasteurization conditions (72°C for 15 s) showed that PC remained almost intact. The CR value of the solutions at pH 5.0, 6.0, and 7.0 was 96%–100%, after incubation at 74°C for 1 min. In addition, PC solution was more stable at low temperature (4°C). Although the CR value of the solutions at room temperature remained higher than 80% after incubation for 10 days, turbidity and odor were observed. Furthermore, the maximum stability at 50°C was observed at pH 6.0, while at 60°C the maximum stability was at pH 5.5. Glucose (20%), sucrose (20%), and sodium chloride (2.5%) were also considered suitable to extend the stability of the PC extract. High temperature at low pH showed to be not effective toward PC stability (Antelo et al., 2008); temperature and

TABLE 6.8   Microalgal Phycobiliprotein Solvent Extraction

Carotenoids	Microalgae Source	Extraction/Purification Method	Yield/Extraction Efficiency	Reference
Allophycocyanin C-phycocyanin	*Spirulina platensis*	100 M phosphate buffer (pH 7.0) at a ratio of 1:100 (w/v) with continuous stirring at 300 rpm at room temperature for 4 h.	N/A	Chaiklahan et al. (2012)
Allophycocyanin C-phycocyanin Phycoerythrin	*Synechococcus* 833	Incubation of sample for 2 h at 37°C, nitrogen cavitation cycles at 1500 psi for 10 min, centrifugation for 40 min at 18,000 rpm to remove cell debris.	85.2%–87.9% DW	Viskari and Colyer (2003)
B-phycoerythrin	*Porphyridium cruentum*	Two-step selective extraction with first passage at 50 MPa in culture medium followed by second passage at 270 MPa in distilled water.	Purity ratio: 0.79	Jubeau et al. (2012)
C-phycocyanin	*Limnothrix* sp.	Distilled water, activated carbon (1% w/v), and chitosan (0.01 g L^{-1}) for extraction.	18% DW	Gantar et al. (2012)
		Ammonium sulfate (25%) for purification at 4°C, overnight. Precipitate resuspended in 0.1 M PBS (pH 7.0) and tangential flow filtration system (30 kDa membrane pore) for pigment concentration.		
C-phycocyanin Phycoerythrin	*Spirulina platensis*	0.1 M PBS at pH of 6.8 and sonication at 28 kHz for extraction. Ultracentrifugation at 200,000g for purification.	Purity: 90%	Furuki et al. (2003)
	Galdieria sulphuraria	PBS at pH of 7.2.	25–30 mg g^{-1}	Sorensen et al. (2013)
	Leptolyngbya sp. KC45	Ammonium sulfate concentration above 1.28 mol L^{-1} ensured only CPC precipitation, together with anion exchange chromatography and tangential flow filtration.	Purity: 3.5–4.5	Pumas et al. (2012)

(Continued)

TABLE 6.8 (*Continued*)   Microalgal Phycobiliproteins Solvent Extraction

Carotenoids	Microalgae Source	Extraction/Purification Method	Yield/Extraction Efficiency	Reference
Phycoerythrin Phycoerythrin	*Leptolyngbya* sp. KC45	Ammonium sulfate at 85% saturation. Purification by three consecutive chromatographic steps; hydroxyapatite column eluted with 100 mM phosphate buffer (pH 7) 0.2 M of NaCl, a Q-sepharose column and a Sephacryl S-200 HR resin.	A565/A280 = 17.3 1.36% yield	Pumas et al. (2012)
	Porphyridium cruentum	Homogenization in 1 M acetic acid–sodium acetate buffer sonication for 10 min, ammonium sulfate precipitation (65% saturation) and dialysis, followed by ion exchange chromatography.	32.7% DW	Bermejo Román et al. (2002)
	Porphyridium cruentum	Cell maceration with glass beads, simultaneous recovery and purification with a (PEG)-phosphate aqueous two-phase system.	A565/A280 = 2.8 76% recovery	Benavides and Rito-Palomares (2006)
	Phormidium sp. A27DM	Freeze-thaw cycles (−30°C and 4°C) in 1 M Tris Cl buffer, two-step ammonium sulfate precipitation at 20% and 70% saturation, and purification by gel permeation chromatography with Sephadex G-150 matrix.	A565/A280 = 3.9 62.6% yield	Parmar et al. (2011)
Phycobiliproteins (APC, CPC, and PE)	*Synechococcus* 833	Buffer: 3% Chaps and 0.3% of olectin for protein solubilization, with cell wall disruption by nitrogen cavitation.	Efficiency: 87.9%	Viskari and Colyer (2003)

pH are indeed inversely proportional with regard to degradation of phycobiliproteins. In addition, at 50°C and 55°C, PC was more stable at pH 6.0, while at 57°C and 65°C the extract was more stable at pH 5.0. Finally, PC solution was stable for a longer period at pH 5.5–6.0. In contrast, pH 5.0 produced a low stability.

ACKNOWLEDGMENTS

A PhD fellowship (Ref. SFRH/BD/62121/2009) for author H.M.A., supervised by author F.X.M. and cosupervised by authors I.S.P. and A.C.G., was granted by Fundação para a Ciência e Tecnologia (FCT, Portugal), under the auspices of ESF and Portuguese funds (MEC). A postdoctoral fellowship (Ref. SFRH/BPD/72777/2010) was granted to author A.C.G, supervised by author F.X.M. and cosupervised by author I.S.P., under the auspices also of ESF and MEC. This research was partially supported by the Strategic Funding UID/Multi/04423/2013 and Project UID/EQU/00511/2013–LEPABE (Laboratory for Process Engineering, Environment, Biotechnology and Energy—EQU/00511) through national funds provided by the FCT and European Regional Development Fund (ERDF), in the framework of program PT2020 and from the European Union (FEDER funds through COMPETE) and national funds (FCT), through the INCENTIVO/EQB/ UI0511/2014 Project.

REFERENCES

Amaro, H.M., Barros, R., Guedes, A.C., Sousa-Pinto, I., Malcata, F.X. Microalgal compounds modulate carcinogenesis in the gastrointestinal tract. *Trends Biotechnol.*, 31 (2013): 92–98.

Antelo, F.S., Costa, J.A.V., and Kalil, S.J. Thermal degradation kinetics of the phycocyanin from *Spirulina platensis. Biochem. Eng. J.*, 41(1) (2008): 43–47.

Aravena, R.I. and del Valle, J. Effect of microalgae preconditioning on supercritical CO_2 extraction of astaxanthin from *Haematococcus pluvialis. 10th International Conference of Supercritical Fluids*, San Francisco, CA, 2012.

Balasundaram, B., Harrison, S., and Bracewell, D.G. Advances in product release strategies and impact on bioprocess design. *Trends Biotechnol.*, 27 (2009): 477–485.

Begum, H., Yusoff, F., Banerjee, S., Khatoon, H., Shariff, M. Availability and utilization of pigments from microalgae. *Crit. Rev. Food Sci. Nutr.*, 2015.

Benavides, J. and Rito-Palomares, M. Bioprocess intensification: A potential aqueous two-phase process for the primary recovery of B-phycoerythrin from *Porphyridium cruentum. J. Chromatogr. B*, 807(1) (2004): 33–38.

Benavides, J. and Rito-Palomares, M. Simplified two-stage method to B-phycoerythrin recovery from *Porphyridium cruentum. J. Chromatogr. B*, 844(1) (2006): 39–44.

Benedetti, S., Benvenuti, F., and Pagliarani, S. Antioxidant properties of a novel phycocyanin extract from the blue-green alga *Aphanizomenon flos-aquae. Life Sci.*, 75(19) (2004): 2353–2362.

Bermejo, R., Alvárez-Pez, J.M., Acién Fernández, F.G., and Molina Grima, E. Recovery of pure B-phycoerythrin from the microalga *Porphyridium cruentum. J. Biotechnol.*, 93(1) (2002): 73–85.

Borowitzka, M.A. High-value products from microalgae—Their development and commercialisation. *J. Appl. Phycol.*, 25(3) (2013): 743–756.

Brunner, G. Supercritical fluids: Technology and application to food processing. *J. Food Eng.*, 67(1–2) (2005): 21–33.

Bryant, D.A., Guglielmi, G., Tandeau de Marsac, N., and Castets, A.M. The structure of cyanobacterial phycobilisomes: A model. *Arch. Microbiol.*, 123(2) (1979): 113–127.

Bustamante, A., Roberts, P., Aravena, R., and del Valle, J.M. Supercritical extraction of astaxanthin from *Haematococcus pluvialis* using ethanol-modified CO_2, experiments and modelling. *11th International Conference of Eng Food*, Athens, Greece, 2011.

Cardoso, L.C., Serrano, C.M., Rodriguez Rodriguez, M., de la Ossa, E.M., and Lubian L.M. Extraction of carotenoids and fatty acids from microalgae using supercritical technology. *Am. J. Anal. Chem.*, 3(12A) (2012): 877–883.

Cerón, M.C., Campos, I.S., Juan, F. et al. Recovery of lutein from microalgae biomass: Development of a process for *Scenedesmus almeriensis* biomass. *J. Agric. Food Chem.*, 56 (2008): 11761–11766.

Cha, K.H., Lee, H.J., Koo, S.Y., Song, D.G., Lee, D.U., and Pan, C.H. Optimization of pressurized liquid extraction of carotenoids and chlorophylls from *Chlorella vulgaris*. *J. Agric. Food Chem.*, 58(2) (2010): 793–797.

Chaiklahan, R., Chirasuwan, N., and Bunnag, B. Stability of phycocyanin extracted from *Spirulina* sp.: Influence of temperature, pH and preservatives. *Process Biochem.*, 47(4) (2012): 659–664.

Chan, M.C., Ho, S.H., Lee, D.J., Chen, C.Y., Huang, C.C., and Chang, J.S. Characterization, extraction and purification of lutein produced by an indigenous microalga *Scenedesmus obliquus* CNW-N. *Biochem. Eng. J.*, 78 (2013): 24–31.

Choi, S.K., Kim, J.H., Park, Y.S., Kim, Y.J., and Chang, H.I. An efficient method for the extraction of astaxanthin from the red yeast *Xanthophyllomyces dendrorhous*. *J. Microbiol. Biotechnol.*, 17(5) (2007): 847–852.

Cubas, C., Gloria Lobo, M., and González, M. Optimization of the extraction of chlorophylls in green beans (*Phaseolus vulgaris* L.) by N,N-dimethylformamide using response surface methodology. *J. Food Compos. Anal.*, 21(2) (2008): 125–133.

Del Campo, A.J., García-González, M., and Guerrero, M.G. Outdoor cultivation of microalgae for carotenoid production: Current state and perspectives. *Appl. Microbiol. Biotechnol.*, 74(6) (2007): 1163–1174.

Denery, J.R, Dragull, K., Tang, C.S., and Li, Q.X. Pressurized fluid extraction of carotenoids from *Haematococcus pluvialis* and *Dunaliella salina* and kavalactones from *Piper methysticum*. *Anal. Chim. Acta*, 501(2) (2004): 175–181.

Devesa, R., Moldes, A., Diaz-Fierros, F., and Barral, M.T. Extraction study of algal pigments in river bed sediments by applying factorial designs. *Talanta*, 72(4) (2007): 1546–1155.

Domínguez-Bocanegra, A.R., Guerrero Legarreta, I., Martinez Jeronimo, F., and Tomasini Campocosio, A. Influence of environmental and nutritional factors in the production of astaxanthin from *Haematococcus pluvialis*. *Bioresour. Technol.*, 92(2) (2004): 209–214.

Dufossé, L., Galaup, P., Yaron, A. et al. Microorganisms and microalgae as sources of pigments for food use: A scientific oddity or an industrial reality? *Trends Food Sci. Technol.*, 16(9) (2005): 389–406.

Eonseon, J., Polle, J.E.W., Lee, H.K., Hyund, S.M., and Chang, M. Xanthophylls in microalgae: From biosynthesis to biotechnological mass production and application. *Microbial Biotechnol.*, 13(2) (2003): 165–174.

Fujii, K. Process integration of supercritical carbon dioxide extraction and acid treatment for astaxanthin extraction from a vegetative microalga. *Food and Bioproducts Processing* 90(4) (2012): 762–766.

Furuki, T., Maeda, S., Imajo, S. et al. Rapid and selective extraction of phycocyanin from *Spirulina platensis* with ultrasonic cell disruption. *J. Appl. Phycol.*, 15(4) (2003): 319–324.

Gantar, M., Simović, D., Djilas, S., Gonzalez, W.W., and Miksovska, J. Isolation, characterization and antioxidative activity of C-phycocyanin from *Limnothrix* sp. strain 37-2-1. *J. Biotechnol.*, 159(1–2) (2012): 21–26.

Glazer, A.N. Phycobiliproteins—A family of valuable widely used fluorophores. *J. Appl. Phycol.*, 6(2) (1994): 105–112.

Goettel, M., Eing, C., Gusbeth, C., Straessner, R., and Frey, W. Pulsed electric field assisted extraction of intracellular valuables from microalgae. *Algal Res.*, 2(4) (2013): 401–408.

Graham, L. and Wilcox, L. *Algae*. Prentice-Hall, Englewood Cliffs, NJ, 2000.

Grimi, N., Dubois, A., Marchal, L., Jubeau, S., Lebovka, N.I., and Vorobiev, E. Selective extraction from microalgae *Nannochloropsis* sp. using different methods of cell disruption. *Bioresour. Technol.*, 153 (2014): 254–259.

Grossman, A.R., Bhaya, D., Apt, K.E., and Kehoe, D.M. Light-harvesting complexes in oxygenic photosynthesis: Diversity, control, and evolution. *Ann. Rev. Genet.*, 29 (1995): 231–288.

Guedes, A.C., Amaro, H.M., and Malcata, F.X. Microalgae as a source of carotenoids. *Mar. Drugs*, 9 (2011b): 625–644.

Guedes, A.C., Amaro, H.M., and Malcata, F.X. Microalgae as sources of high added-value compounds—A brief review of recent work. *Biotechnol. Progress*, 27 (2011a): 597–613.

Guedes, A.C., Gião, M.S., Matias, A.A., Nunes, A.V.M., Pintado, M.E., Duarte, C.M.M, and Malcata, F.X. Supercritical fluid extraction of carotenoids and chlorophylls a, b and c, from a wild strain of *Scenedesmus obliquus* for use in food processing. *J. Food Eng.*, 116(2) (2013): 478–482.

Guedes, A.C. and Malcata, F.X. Chapter 4—Nutritional value and uses of microalgae in aquaculture. In: *Aquaculture*. Muchlisin, Z.A., Ed. InTech, Rijeka, Croatia, 2012, ISBN: 978-953-307-974-5.

Hejazi, M.A., De Lamarliere, C., Rocha, J.M.S., Vermuë, M., Tramper, J., and Wijffels, R.H. Selective extraction of carotenoids from the microalga *Dunaliella salina* with retention of viability. *Biotechnol. Bioeng.*, 79(1) (2002): 29–36.

Herrero, M., Martin-Alvarez, P.J., Senorans, F.J., Cifuentes, A., and Ibanez, E. Optimization of accelerated solvent extraction of antioxidants from *Spirulina platensis* microalga. *Food Chem.*, 93(3) (2005): 417–423.

Herrero, M., Mendiola, J.A., Cifuentes, A., and Ibáñez, E. Supercritical fluid extraction: Recent advances and applications. *J. Chromatogr. A*, 1217(16) (2010): 2495–2511.

Hosikian, A., Lim, S., Halim, R., and Danquah, M.K. Chlorophyll extraction from microalgae: A review on the process engineering aspects. *Int. J. Chem. Eng.*, (2010): 11.

Hu, C.C., Lin, J.T., Lu, F.L., Chou, F.P., and Yang, D.J. Determination of carotenoids in *Dunaliella salina* cultivated in Taiwan and antioxidant capacity of the algal carotenoid extract. *Food Chem.*, 109(2) (2008): 439–446.

Jaime, L., Rodriguez-Meizoso, I., Cifuentes, A., Santoyo, S., Suarez, S., Ibanez, E., and Javier, S.F. Pressurized liquids as an alternative process to antioxidant carotenoids' extraction from *Haematococcus pluvialis* microalgae. *LWT—Food Sci. Technol.*, 43(1) (2010): 105–112.

Jeffrey, S.W. Quantitative thin-layer chromatography of chlorophylls and carotenoids from marine algae. *Biochim. Biophys. Acta*, 162(2) (1968): 271–285.

Jeffrey, S.W., Mantoura, R.F.C., and Wright, S.W., Eds. *Phytoplankton Pigments in Oceanography: Guidelines to Modern Methods*. UNESCO, Paris, France, 1997.

Jubeau, S., Marchal, L., Pruvost, J., Jaouen, P., Legrand, J., and Fleurence, J. High pressure disruption: A two-step treatment for selective extraction of intracellular components from the microalga *Porphyridium cruentum*. *J. Appl. Phycol.*, 25(4) (2012): 983–989.

Jyonouchi, H., Sun, S., and Gross, M. Effect of carotenoids on in vitro immunoglobulin production by human peripheral blood mononuclear cells: Astaxanthin, a carotenoid without vitamin A activity, enhances in vitro immunoglobulin production in response to a T-dependent stimulant and antigen. *Nutr. Cancer*, 23(2) (1994): 171–183.

Kang, C.D. and Sim, S.J. Selective extraction of free astaxanthin from *Haematococcus* culture using a tandem organic solvent system. *Biotechnol. Progress*, 23(4) (2007): 866–871.

Kawsar, S.M.A., Fujii, Y., Matsumoto, R., Yasumitsu, H., and Ozeki, Y. Protein R-phycoerythrin from marine red alga *Amphiroa anceps*: Extraction, purification and characterization. *Phytologia Balcanica*, 17(3) (2011): 347–354.

Kitada, K., Machmudah, S., Sasaki, M., Goto, M., Nakashima, Y., Kumamoto, S., and Hasegawa, T. Supercritical CO_2 extraction of pigment components with pharmaceutical importance from *Chlorella vulgaris*. *J. Chem. Technol. Biotechnol.*, 84(5) (2009): 657–661.

Koo, S.Y., Cha, K.H., Song, D.G., Chung, D., and Pan, C.H. Optimization of pressurized liquid extraction of zeaxanthin from *Chlorella ellipsoidea*. *J. Appl. Phycol.*, 24(4) (2012): 4725–730.

Krichnavaruk, S., Shotipruk, A., Goto, M., and Pavasant, P. Supercritical carbon dioxide extraction of astaxanthin from *Haematococcus pluvialis* with vegetable oils as co-solvent. *Bioresour. Technol.*, 99(13) (2008): 5556–5560.

Lee, R.E. *Phycology*. Cambridge University Press, UK, 1999.

Liu, Y., Xu, L., and Cheng, N. Inhibitory effect of phycocyanin from *Spirulina platensis* on the growth of human leukemia k562 cells. *J. Appl. Phycol.*, 12(2) (2000): 125–130.

Luengo, E., Condón-Abanto, S., Álvarez, I., and Raso, J. Effect of pulsed electric field treatments on permeabilization and extraction of pigments from *Chlorella vulgaris*. *J. Membrane Biol.*, 247(12) (2014): 1269–1277.

Luengo, E., Martínez, J.M., Bordetas, A., Álvarez, I., and Raso, J. Influence of the treatment medium temperature on lutein extraction assisted by pulsed electric fields from *Chlorella vulgaris*. *Innovat. Food Sci. Emerg. Technol.*, (2015): 15–20.

Lynn Co, D.Y.C. and Schanderl, S.H. Separation of chlorophylls and related plant pigments by two-dimensional thin-layer chromatography. *J. Chromatogr. A*, 26 (1967): 442–448.

Ma, R.Y.-N. and Chen, F. Enhanced production of free trans-astaxanthin by oxidative stress in the cultures of the green microalga *Chlorococcum* sp. *Process Biochem.*, 36(12) (2001): 1175–1179.

Machmudah, S., Shotipruk, A., Goto, M., Sasaki, M., and Hirose, T. Extraction of astaxanthin from *Haematococcus pluvialis* using supercritical CO_2 and ethanol as entrainer. *Ind. Eng. Chem. Res.*, 45(10) (2006): 3652–3657.

Macías-Sánchez, M.D., Fernandez-Sevilla, J.M., Fernández, F.G.A., García, M.C.C., and Grima, E.M. Supercritical fluid extraction of carotenoids from *Scenedesmus almeriensis*. *Food Chem.*, 123(3) (2010): 928–935.

Macías-Sánchez, M.D., Mantell, C., Rodríguez, M., de La Ossa, E.M., Lubián, L.M., and Montero, O. Supercritical fluid extraction of carotenoids and chlorophyll *a* from *Nannochloropsis gaditana*. *J. Food Eng.*, 66(2) (2005): 245–251.

Macías-Sánchez, M.D., Mantell, C., Rodríguez, M., de la Ossa, E.M., Lubián, L.M., and Montero, O. Comparison of supercritical fluid and ultrasound-assisted extraction of carotenoids and chlorophyll *a* from *Dunaliella salina*. *Talanta*, 77(3) (2009a): 948–952.

Macías-Sánchez, M.D., Serrano, C.M., Rodríguez, M., Rodríguez, M.R., and de la Ossa, M.E. Kinetics of the supercritical fluid extraction of carotenoids from microalgae with CO_2 and ethanol as co-solvent. *Chem. Eng. J.*, 150(1) (2009b): 104–113.

Madgwick, J.C. Chromatographic determination of chlorophylls in algal cultures and phytoplankton. *Deep Sea Res. Oceanogr. Abstr.*, 13(3) (1966): 459–466.

Mäki-Arvela, P., Hachemi, I., and Murzin, D.Y. Comparative study of the extraction methods for recovery of carotenoids from algae: Extraction kinetics and effect of different extraction parameters. *J. Chem. Technol. Biotechnol.*, 89(11) (2014): 1607–1626.

Mantoura, R.F.C. and Llewellyn, C.A. The rapid determination of algal chlorophyll and carotenoid pigments and their breakdown products in natural waters by reverse-phase high-performance liquid chromatography. *Anal. Chim. Acta*, 151(2) (1983): 297–314.

Mendes, R.L., Nobre, B.P., Cardoso, M.T., Pereira, A.P., and Palavra, A.F. Supercritical carbon dioxide extraction of compounds with pharmaceutical importance from microalgae. *Inorgan. Chim. Acta*, 356 (2003): 328–334.

Nagaoka, S., Shimizu, K., Kaneko, H., Shibayama, F., Morikawa, K., Kanamaru, Y., Otsuka, A., Hirahashi, T., and Kato, T. A novel protein C phycocyanin plays a crucial role in the hypocholesterolemic action of *Spirulina platensis* concentrate in rats. *J. Nutr.*, 135(10) (2005): 2425–2430.

Orosa, M., Franqueira, D., Cid, A., and Abalde, J. Analysis and enhancement of astaxanthin accumulation in *Haematococcus pluvialis*. *Bioresour. Technol.*, 96(3) (2005): 373–378.

Parmar, A., Singh, N.K., Kaushal, A., and Madamwar, D. Characterization of an intact phycoerythrin and its cleaved 14kDa functional subunit from marine cyanobacterium *Phormidium* sp. A27DM. *Process Biochem.*, 46(9) (2011): 1793–1799.

Parniakov, O., Barba, F.J., Grimi, N., Marchal, L., Jubeau, S., Lebovka, N., and Vorobiev, E. Pulsed electric field assisted extraction of nutritionally valuable compounds from microalgae *Nannochloropsis* spp. using the binary mixture of organic solvents and water. *Innovat. Food Sci. Emerg. Technol.*, 27 (2015): 79–85.

Pasquet, V., Chérouvrier, J.-R., Farhat, F., Thiérya, V., Piota, J.-M., Bérardb, J.-B., Kaasb, R., Seriveb, B., Patricec, T., Cadoretb, J.-P., and Picot, L. Study on the microalgal pigments extraction process: Performance of microwave assisted extraction. *Process Biochem.*, 46(1) (2011): 59–67.

Plaza, M., Cifuentes, A., and Ibáñez, E. In the search of new functional food ingredients from algae. *Trends. Food. Sci. Tech.*, 19(1) (2008): 31–39.

Plaza, M., Santoyo, S., Jaime, L., Garcia-Blairsy, R.G., Herrero, M., Senorans, F.J., Ibanez, E. Screening for bioactive compounds from algae. *J. Pharm. Biomed. Anal.*, 51(2) (2010): 450–455.

Porra, R.J. Recent advances and re-assessments in chlorophyll extraction and assay procedures for terrestrial, aquatic, and marine organisms, including recalcitrant algae. In: *Chlorophylls*. Scheer, H., Ed. CRC Press, Boca Raton, FL, 1991, pp. 31–57.

Prabakaran, P. and Ravindran, A.D. A comparative study on effective cell disruption methods for lipid extraction from microalgae. *Lett. Appl. Microbiol.*, 53(2) (2011): 150–154.

Pumas, C., Peerapornpisal, Y., Vacharapiyasophon, P., Leelapornpisid, P., Boonchum, W., Ishii, M., and Khanongnuch, C. Purification and characterization of a thermostable phycoerythrin from hot spring cyanobacterium *Leptolyngbya* sp. KC45. *Int. J. Agric. Biol.*, 14(1) (2012): 121–125.

Reyes, F.A., Mendiola, J.A., Ibañez, E., and del Valle, J.M. Astaxanthin extraction from *Haematococcus pluvialis* using CO_2-expanded ethanol. *J. Supercrit. Fluids*, 92 (2014): 75–83.

Riley, J.P and Wilson, T.R.S. Use of thin-layer chromatography for separation and identification of phytoplankton pigments. *J. Mar. Biol. Assoc. U.K.*, 45(3) (1965): 583–591.

Rodríguez-Meizoso, I., Jaime, L., Santoyo, S., Cifuentes, A., Garcia-Blairsy, R.G., Senorans, F.J., and Ibanez, E. Pressurized fluid extraction of bioactive compounds from *Phormidium* species. *J. Agric. Food Chem.*, 56(10) (2008): 3517–3523.

Ruen-ngam, D., Shotipruk, A., Pavasant, P., Machmudah, S., and Goto, M. Selective extraction of lutein from alcohol treated *Chlorella vulgaris* by supercritical CO_2. *Chem. Eng. Technol.*, 35(2) (2012): 255–260.

Sahena, F., Zaidul, I.S.M., Jinap, S., Karimb, A.A., Abbasa, K.A., Norulainic N.A.N., and Omar, A.K.M. Application of supercritical CO_2 in lipid extraction—A review. *J. Food Eng.*, 95(2) (2009): 240–253.

Sang, M.K., Jung, Y.J., Kwon, O.N., Cha, K.H., Um, B.H., Chung, D., and Pan, C.H. A potential commercial source of fucoxanthin extracted from the microalga *Phaeodactylum tricornutum*. *Appl. Biochem. Biotechnol.*, 166(7) (2012): 1843–1855.

Santiago-Santos, MaC., Ponce-Noyola, T., Olvera-Ramírez, R., Ortega-López, J., Cañizares Villanueva, R.O. Extraction and purification of phycocyanin from *Calothrix* sp. *Process Biochem.*, 39(12) (2004): 2047–2052.

Sarada, R., Tripathi, U., and Ravishankar, G. Influence of stress on astaxanthin production in *Haematococcus pluvialis* grown under different culture conditions. *Process Biochem.*, 37(6) (2002): 623–627.

Sartory, D.P. The determination of algal chlorophyllous pigments by high performance liquid chromatography and spectrophotometry. *Water Res.*, 19(5) (1985): 605–610.

Sartory, D.P. and Grobbelaar, J.U. Extraction of chlorophyll a from freshwater phytoplankton for spectrophotometric analysis. *Hydrobiologia*, 114(3) (1984): 177–187.

Scheer, H., William, J.L., and Lane, M.D. *Encyclopedia of Biological Chemistry: Chlorophylls and Carotenoids*. Elsevier, New York, 2004, pp. 430–437.

Schumann, R., Häubner, N., Klausch, S., and Karsten, U. Chlorophyll extraction methods for the quantification of green microalgae colonizing building facades. *Int. Biodeterior. Biodegr.*, 55(3) (2005): 213–222.

Sekar, S. and Chandramohan, M. Phycobiliprotein as a commodity: Trends in applied research, patents and commercialization. *J. Appl. Phycol.*, 20(2) (2008): 113–136.

Serive, B., Kaas, R., Bérard, J.-B., Pasquet, V., Picot, L., and Cadoret, J.-P. Selection and optimisation of a method for efficient metabolites extraction from microalgae. *Bioresour. Technol.*, 124 (2012): 311–320.

Shibata, S., Ishihara, C., and Matsumoto, K. Improved separation method for highly purified lutein from *Chlorella* powder using jet mill and flash column chromatography on silica gel. *J. Agric. Food Chem.*, 52(20) (2004): 6283–6286.

Shoaf, W.T. Rapid method for the separation of chlorophylls *a* and *b* by high-pressure liquid chromatography. *J. Chromatogr. A*, 152(1) (1978): 247–249.

Simon, D. and Helliwell, S. Extraction and quantification of chlorophyll *a* from freshwater green algae. *Water Res.*, 32(7) (1998): 2220–2223.

Sørensen, L., Hantke, A., and Eriksen, N.T. Purification of the photosynthetic pigment C-phycocyanin from heterotrophic *Galdieria sulphuraria*. *J. Sci. Food Agric.*, 93(12) (2013): 2933–2938.

Szymczak-Żyła, M., Kowalewska, G., and Louda, J.W. Chlorophyll-*a* and derivatives in recent sediments as indicators of productivity and depositional conditions. *Mar. Chem.*, 125(1–4) (2011): 39–48.

Talisic, G.C., Yumang, A.N, and Salta, M.T.S. Supercritical fluid extraction of β-carotene from *D. salina* algae using C_2H_6 and C_2H_2. *Int. Conf. Geol. Environ. Sci.*, Singapore, 36 (2012): 30–34.

Van Den Hoek, C., Mann, D.G., and Jahns, H.M. *Algae: An Introduction of Phycology*, Cambridge University Press, Cambridge, UK, 1995.

Vanthoor-Koopmans, M., Wijffels, R.H., Barbosa, M.J., and Eppink, M.H.M. Biorefinery of microalgae for food and fuel. *Bioresour. Technol.*, 135 (2013): 142–149.

Viskari, P.J. and Colyer, C.L. Rapid extraction of phycobiliproteins from cultured cyanobacteria samples. *Anal. Biochem.*, 319(4) (2003): 263–271.

Wang, L. and Weller, C.L. Recent advances in extraction of nutraceuticals from plants. *Trends Food Sci. Technol.*, 17(6) (2006): 300–312.

Wright, S.W. and Shearer, J.D. Rapid extraction and high-performance liquid chromatography of chlorophylls and carotenoids from marine phytoplankton. *J. Chromatogr. A*, 294 (1984): 281–295.

Yen, H.W., Chiang, W.C., and Sun, C.H. Supercritical fluid extraction of lutein from *Scenedesmus* cultured in an autotrophical photobioreactor. *J. Taiwan Inst. Chem. Eng.*, 43(1) (2012): 53–57.

Zbinden, M.D.A., Sturm, B.S.M., Nord, R.D., Carey, W.J., Moore, D., Shinogle, H., and Stagg-Williams, S.M. Pulsed electric field (PEF) as an intensification pre-treatment for greener solvent lipid extraction from microalgae. *Biotechnol. Bioeng.*, 110(6) (2013): 1605–1615.

Zhao, L., Zhao, G., Chen, F., Wang, Z., Wu, J., and Hu, X. Different effects of microwave and ultrasound on the stability of (all-E)-astaxanthin. *J. Agric. Food Chem.*, 54(21) (2006): 8346–8351.

CHAPTER 7

Opening Avenues in Marine Probiotics
Present and Future

Ira Bhatnagar, Mani Vasagan, and P. V. Bramhachari

CONTENTS

7.1 Introduction: An Ode to the Saga   129
7.2 Nitty-Gritty of Probiotics   131
7.3 Probiotics in Aquaculture: Gone Fishy   131
   7.3.1 Probiotic Preparation   132
   7.3.2 Probiotic Strains Studied in Aquaculture   133
   7.3.3 Fish Eggs and Larvae   133
   7.3.4 Fish Juveniles and Adults   134
   7.3.5 Crustaceans   134
   7.3.6 Mollusks   135
7.4 Probiotics in Larviculture: Rearing Life   135
7.5 Probiotics in Public Health: Rejuvenating Fitness   136
7.6 Marine Probiotics: Approaches for Development   136
7.7 Nanotechnology Applications in Aquaculture   137
7.8 Promises and Hopes   138
7.9 Future Directions: Ray of Hope   139
References   139

7.1 INTRODUCTION: AN ODE TO THE SAGA

The concept of probiotics (which means, "for life") was introduced in the early twentieth century by Elie Metchnikoff (1907), but it remained mostly unknown to date and received considerable attention in the recent past with the advent of functional and health food market globally. Technically speaking, probiotics are "good bacteria" that are similar to the ones found in our gut. They facilitate our gut by increasing the number of good bacteria and, in turn, inhibit the bad ones. On the other hand, *prebiotics* act as the food for probiotic bacteria where they stimulate their growth and activity. When pre- and probiotics are mixed together, they form *symbiotics*.

Gatesoupe (1999) defines probiotics as "live microbial cells administered to cultured organisms to colonize the digestive tract and improve their immune response." Probiotics are useful in a number of ways as they are highly indispensable for the proper development of our immune system. Despite this, probiotics protect us from disease-causing micro-organisms and are vital for the digestions of nutrients and food. Indeed there have been a number of studies that suggest that probiotics ease symptoms of lactose intolerance and irritable bowel syndrome. Some have also been shown to relieve constipation and diarrhea, as well as reduce the risk of osteoporosis as they are believed to augment the assimilation of calcium and magnesium.

Researchers are now trying to explore the possibilities of therapeutic applications of probiotics in inflammatory bowel disease, treating diarrhea, eczema prevention in children, and reducing bladder cancer recurrence and urinary tract infections. It is worth mentioning here that our knowledge of probiotics should include the fact that the beneficiary effects of probiotics vary from individual to individual depending upon how the immune system responds to it. Further, the health benefits depend on the particular strain of the organism used as a probiotic and the dose. For example, the best known probiotic strains are *Bifidobacteria*, *Lactobacilli*, and *Streptococcus thermophilus* and can be found in food products such as yogurts, fermented and unfermented milk, miso, tempeh, and some juices and soy beverages. The use of probiotics is increasing day by day as they are easy to consume and can also be obtained in the form of dietary supplements such as capsules or powders. In fact, certain foods that enhance the probiotic effect, known as prebiotics, are found in whole grains, almonds, bananas, garlic, honey, leeks, onions, and artichokes, which makes it even more easy to lift up the health benefits of probiotic strains.

As mentioned earlier, probiotics act in a two way fashion, either by enhancing the growth of beneficial bacteria or by reducing the growth of harmful germs and microbes. This property of probiotic strains can be explored to produce functional foods. This property may also be helpful in food storage industry as prolonged storage of foods under appropriate conditions leads to the development of decaying bacteria. Probiotic strains can well be used in aquaculture for proper growth and maintenance of the cultured species. The microbiota in the aquatic environment is usually in equilibrium, and it is composed by bacteria that are either benefic or neutral to cultured animals, or by harmful obligate and opportunistic pathogenic bacteria (Schulze et al., 2006). Owing to inadequate maintenance, pathogenic bacteria may proliferate, disturbing the equilibrium (Karunasagar et al., 1994) and incurring further financial loss due to decay of the economically important species. As a general basis, antibiotics are basically used to encounter the bacterial disease in marine shrimp culture. However, their indiscriminate use has led to the appearance of antibiotic-resistant strains (Skjermo and Vadstein, 1999) and contamination of shrimp meat and also the environment (Holmström et al., 2003). Therefore, several countries have banned the use of antibiotics such as chloramphenicol (FAO, 2002). Thus, as a substitute to the use of antibiotics, the dietary supplementation with probiotic bacteria is being widely employed in the aquaculture industry (Gatesoupe, 1999; Vine et al., 2006). Among the probiotic bacteria used in aquaculture, the lactic acid bacteria stand out for their easy multiplication, production of antimicrobial compounds (bacteriocins, hydrogen peroxide, organic, and lactic acids), and the stimulation of the nonspecific immune response of the host (Gatesoupe, 2008). Studies have demonstrated the beneficial effect of the addition of such bacteria in the culture of several aquatic species. Nevertheless, only a few studies have been conducted to assess the use of lactic acid bacteria strains isolated from marine shrimp (Vieira et al., 2010). This chapter is an attempt to bring forward the beneficial effects of probiotics and the prerequisites for a strain to be considered as a probiotic and explore the possibilities of developing marine probiotics.

7.2 NITTY-GRITTY OF PROBIOTICS

Probiotic strains should meet some standard criterion to be considered safe for use. As mentioned earlier, probiotics are bacteria that help maintain the natural balance of microflora in the intestines. The normal human digestive tract contains about 400 types of probiotic bacteria that reduce the growth of harmful bacteria and promote a healthy digestive system. A number of reports from the past suggest experimenting on the benefits of probiotic therapies and propose an array of potentially advantageous medicinal uses for probiotics. However, since it's a matter of safeguarding living creatures, various concerns exist regarding the issues of safety, efficacy, and reliability, as well as the labeling of the probiotic products being formulated. To ensure strict compliance to the set parameters for probiotic use and safety, it is mandatory to define certain guidelines that should be followed for a product/strain to be termed as probiotic and before a potential probiotic strain enters the market for further use. It is also necessary to follow the clinically relevant steps to move probiotics closer to being embraced by the medical community. The first and foremost step for the same is the proper identification of stain and in vitro screening for its probiotic characteristics. Second, before its official use, animal studies should be rigorously pursued to establish safety and in vivo animal and human studies to establish efficacy. Phase I, II, and III clinical trials to prove health benefits that are as good as or better than standard prevention or treatments for a particular condition or disease should be followed. The optimism associated with probiotics is, however, counterbalanced by skepticism as many "probiotic" products in the market are unreliable in content and unproven clinically (Hughes and Hillier, 1990; Zhong et al., 1998; Hamilton-Miller et al., 1999). Following good manufacturing practices and production of high quality products are also very important in the probiotic field. Incongruously labeled probiotic may also lead to chaos and inapt use of the same. To avoid misleading the consumers and further damage to the aquaculture, informative/precise labeling of probiotic products is required, which should include information about the strain, viable numbers at the end of shelf life, and adequate storage conditions. Other than this, some further steps include studies to identify the mechanism of action of the studied probiotic in vivo; development of probiotic organisms that can carry vaccines to hosts and/or antiviral probiotic; expansion of proven strains to benefit the oral cavity, nasopharynx, respiratory tract, stomach, vagina, bladder, and skin; and for cancer, allergies, and recovery from surgery/injury (Reid, 2005). These guidelines and steps should be strictly followed to ensure a high quality, effective probiotic with health benefits.

7.3 PROBIOTICS IN AQUACULTURE: GONE FISHY

In aquaculture, exposure to a diverse bacterial microflora is limited by the resources available. Therefore, the gastrointestinal flora usually resembles the microflora initially present in the rearing water, microalgae, and live food. Aquaculture has become an important economic activity in many countries. In large-scale production facilities, where aquatic animals are exposed to stressful conditions, problems related to diseases and deterioration of environmental conditions often occur and result in serious economic losses (Bondad-Reantaso et al., 2005). In recent decades, disease prevention and control have led to a substantial increase in the use of chemical additives and veterinary medicines. The utility of antimicrobial agents as a preventive measure has been questioned, given extensive documentation of the evolution of antimicrobial resistance among pathogenic bacteria (Nomoto, 2005). In addition, there are environmental problems associated with the chemical additives (Wang and Xu, 2004). Therefore, the need for alternative techniques is increasing and the contribution of probiotics may be considerable. The word "probiotic" was introduced by Parker (1974). According to his original definition, probiotics are "organisms and substances which contribute to intestinal microbial balance" (Parker, 1974). Fuller (1989) revised the definition as "live microbial feed supplement which beneficially affects the host animal by improving its intestinal microbial balance" (Fuller, 1989). Therefore, several terms such as "friendly," "beneficial," or "healthy" bacteria are also commonly used to describe probiotics. Although application of probiotics in aquaculture seems to be relatively recent (Kozasa, 1986), the interest in such environment friendly treatments is increasing rapidly. Moriarty (1998) proposed to extend the definition of probiotics in aquaculture to microbial "water additives" (Moriarty, 1998). A growing number of studies have dealt explicitly with probiotics, and it is now possible to survey its state of the art, from the empirical use to the scientific approach (Gatesoupe, 1991; Vine et al., 2006; Wang, 2007). The probiotic use and disease management in aquaculture with special reference to molluscan culture have been discussed and reviewed (Kesarcodi-Watson et al., 2008). However, there has been little review of the rationale from the perspective of the digestive tract and the safety evaluation of probiotics in aquaculture. The use of microbial probiotics in aquaculture is now widely accepted (Mohanty et al., 1993, 1996; Gatesoupe, 1999; Gomez-Gil et al., 2000; Verschuere et al., 2000; Irianto and Austin, 2002; Vine et al., 2006; Yanbo and Zirong, 2006; Wang, 2007; Sharma and Bhukar, 2011). Nowadays, a number of preparations of probiotics are commercially available and have been introduced to fish, shrimp, and molluscan farming as feed additives or are incorporated in pond water (Wang et al., 2005). According to the claims of the producers, these products are effective in supporting the health of aquatic animals and are also safe. On the other hand, there are doubts with regard to the general concept of probiotics and to these claims. Thus, there is clearly a need to increase our knowledge of intestinal microbiology and the effective preparation and safety evaluation of probiotics.

7.3.1 Probiotic Preparation

Although specific numbers are not mentioned in the definition, high levels of viable microorganisms are recommended in probiotics for efficacy (Gatesoupe, 1999). Consequently, the retention of high viability during preparation and storage presents particular challenges and can be regarded as a major bottleneck in commercial probiotic production. This is particularly the case for "technologically sensitive" strains (e.g., some lab species) with the result that most successfully marketed probiotics are usually robust only in nature. Most liquid/frozen probiotic cultures require refrigeration for storage and distribution, thereby adding expense and inconvenience to their widespread use in aquaculture. The survival and viable cell count of probiotic bacteria vary depending on the strains and manufacturers (Schillinger, 1999). To maintain confidence in probiotic products used in aquaculture, it is important to demonstrate good survival of the bacteria in products during their shelf life. Over the past 20 years, the understanding of how to handle and prepare probiotic products has grown. The majority of probiotic studies have been conducted in vitro and often without the use of homogeneous standards.

The genus *Bacillus* is Gram-positive rods that form a single endospore (spore) and represent a peculiar case among the bacteria used as probiotics. *Bacillus* spp., which include *B. subtilis*, *B. cereus*, *B. coagulans*, *B. clausii*, *B. megaterium*, and *B. licheniformis*, are used as probiotics (Oggioni et al., 2003). Due to the physical and biological characteristics of the spore, these preparations (powders or spore suspensions in distilled water) are extremely resistant to the environment and have a prolonged shelf life. Also, the cost of production of spores for aquaculture is low with respect to the production of purified components. The option to use these organisms as probiotics is made even more feasible by the well-described systems, including the specific plasmids available for genetic engineering of *B. subtilis* (Driks, 1999). However, as a general rule, the presence of plasmids is not a reason to discard the strain as a potential probiotic, but the role of this extrachromosomal DNA in establishing phenotypes relevant to technological and probiotic properties must be assessed.

Stability is also critical to guarantee the efficacy of probiotics and their ability to induce the beneficial effects in the host during final product formulation (Fonseca et al., 2001). Therefore, to be effective and confer these health benefits, probiotic cultures must be able to retain their probiotic properties after processing and with sufficient numbers survive during shelf life/storage. It is well known that the stability of probiotics is influenced by various factors, including the species, strain biotype, water activity, temperature, hydrogen ion concentration (pH), osmotic pressure, mechanical friction, and oxygen. Consequently, special attention and techniques are needed during the process of probiotic production. Different approaches that increase the resistance of these probiotics against adverse conditions have been proposed, including use of oxygen-impermeable containers, two-step fermentation, stress adaptation, incorporation of micronutrients such as peptides and amino acids, and microencapsulation (Gismondo et al., 1999).

7.3.2 Probiotic Strains Studied in Aquaculture

Most probiotics proposed as biological control agents in aquaculture belong to lactic acid bacteria (*Lactobacillus* and *Carnobacterium*), to the genus *Vibrio* (*V. alginolyticus*), to the genus *Bacillus*, or to the genus *Pseudomonas*, although other genera or species have also been mentioned (*Aeromonas* and *Flavobacterium*).

7.3.3 Fish Eggs and Larvae

The spawning and early life stages of fish larvae may have profound implications for the dynamics of microbial communities. These communities can also be influenced by inorganic and organic compounds. Immediately after the hatching process, fish larvae come into contact with their immediate environment that provides colonization by a wide variety of microorganism. It is therefore evident that health status will depend on management conditions to constitute the egg microbiota, an important factor in the first establishment of an indigenous microbiota. Viable counts from water in the hatching units have been on the order of 10^3 mL^{-1} prior to hatching and 10^6 mL^{-1} 2 days after hatching during early life of fish larvae in intensive rearing systems (Hansen and Olafsen, 1989). The composition of microbiota is influenced by many factors including the availability of nutrients, animal physiology, and immunological factors. Under normal conditions, one of the basic physiological functions of the resident microbiota is that it functions as a microbial barrier against microbial pathogens and as a complement to the establishment of digestive enzymes. During the initial feeding period, it is possible to manipulate the establishment of an artificial dominance of a determined group of bacteria in the fish-associated microbiota by adding a specific strain. A significant increase in the mean weight and survival rate of turbot larvae (*S. maximus*) fed rotifers enriched in lactic acid bacteria, and these strains provided a significant protection against a pathogenic *Vibrio* compared to control larvae (Gatesoupe, 1994). Similarly, the addition of *Carnobacterium divergens* showed a certain improvement in disease resistance in cod (*Gadus morhua*) fry after a challenge with *V. anguillarum* (Gildberg and Mikkelsen, 1998). In a study in the rearing of larvae turbot, 34 strains of 400 marine bacteria exhibited in vitro antibacterial activity against *V. anguillarum*, *Vibrio splendidus*, and *Pseudoalteromonas*. These strains were identified as *Roseobacter* spp., *Vibrio* spp., and *Pseudoalteromonas*. *Roseobacter* spp. were not lethal to egg yolk sac turbot larvae, and in two of three trials, the mortality of larvae decreased in treatments where 10^7 cfu mL^{-1} *Roseobacter* sp. strain 27-4 was applied (Hjelm et al., 2004).

7.3.4 Fish Juveniles and Adults

In experiments performed by Queiroz and Boyd (2007), it was reported that a commercially prepared bacterial mixture of *Bacillus* spp. mixed into the rearing water increased survival and production of channel catfish (*Ictalurus punctatus*) (Queiroz and Boyd, 2007). A strain of *Carnobacterium* sp. previously isolated from the intestine of Atlantic salmon was effective at controlling infections caused by *A. salmonicida*, *Vibrio ordalii*, and *Yersinia ruckeri* in fry and fingerling salmonids, applied at 5×10^7 cells g^{-1} of feed (Robertson et al., 2000). Cultures of *Aeromonas hydrophila*, *Vibrio fluvialis*, *Carnobacterium* sp., and an unidentified Gram-positive coccus have been beneficial for rainbow trout when these strains were administered as food additives, since their application reduced significantly the impact of furunculosis by competitive exclusion and enhanced cellular immunity in the fish (Irianto and Austin, 2002). Similarly, a *V. alginolyticus* strain at 10^8 cells mL^{-1} was applied in a bath treatment to Atlantic salmon. Experiments revealed that application of the probiotic led to a reduction in mortality after exposures to *A. salmonicida* and to a lesser extent after exposures to *V. anguillarum* and *V. ordalii* (Austin et al., 2006). Recently, studies have reported the presence of antifungal effects from various strains of probiotics. For example, a strain isolated from freshwater, *Aeromonas media* (strain A199) in culture of eels (*Anguilla australis* Richardson), presented antagonistic activity against *Saprolegnia* sp., suppressing the growth of this opportunistic pathogen (Lategan and Gibson, 2003).

7.3.5 Crustaceans

In a study of tiger shrimp, the inoculation of *Bacillus* S11, a saprophytic strain, resulted in greater survival of the postlarval *P. monodon* that were challenged by pathogenic luminescent bacterial culture (Rengpipat et al., 1998). A mixture of *Lactobacillus* spp. isolated from chicken gastrointestinal tracts has improved the growth and survival rates of juvenile *P. monodon* when fed these strains for 100 days (Phianphak et al., 1999). Recently, the growth of pathogenic *V. harveyi* was controlled by the probiotic effect of *Bacillus subtilis* BT23 under in vitro and in vivo conditions. Improved disease resistance was observed after exposing juvenile *P. monodon* to *B. subtilis* BT23, isolated from shrimp culture ponds, at a density of 10^6–10^8 cells mL^{-1}, for 6 days before a challenge with *V. harveyi* at 10^3–10^4 cells mL^{-1} for 1 h infection with a 90% reduction in accumulated mortality (Vaseeharan and Ramasamy, 2003). The probiotic effect in *L. vannamei* has been reported using three strains isolated from the hepatopancreas of shrimp. These strains were identified as *Vibrio* P62, *Vibrio* P63, and *Bacillus* P64 and achieved inhibition percentages against *V. harveyi* S2 under in vivo conditions of 83%, 60%, and 58%, respectively. Histologic analyses after the colonization and interaction experiment confirmed that the probiotic strains had no pathogenic effect on the host (Gullian et al., 2004). Also, *Pseudomonas* sp. PM 11 and *V. fluvialis* PM 17 have been selected as candidate probiotics isolated from the gut of farm reared tiger shrimp by the ability to secrete extracellular macromolecule digesting enzymes. However, when shrimps were treated with each of the candidate strains, the estimation of immunological indicators such as hemocyte counts, phenol oxidase, and antibacterial activity showed declining trends (Gullian et al., 2004). Possibly these bacteria did not colonize the gut; therefore, they did not help in improving the immune system of shrimp. It is known that colonization with specific microbiota in the gut may play a role in balancing the intestinal mucosal immune system, which may contribute to the induction and maintenance of immunological tolerance or to the inhibition of the deregulated responses induced by pathogens in host.

7.3.6 Mollusks

Studies in mollusks have determined that *Roseobacter* sp. (strain BS107) in coculture with *V. anguillarum* (strain 408) displayed an inhibitory effect on *Vibrio*, enhancing the survival of scallop (*Pecten maximus*) larvae (Ruiz-Ponte et al., 1999). Similarly, cultures of *A. media* have controlled infections by *Vibrio tubiashii* in oyster (*Crassostrea gigas*) larvae (Gibson et al., 1998). Recently, the use of *Alteromonas haloplanktis* (strain 77) and *Vibrio* sp. (strain 11) has been effective at controlling infections by *V. anguillarum* in scallop (*Argopecten purpuratus*) larvae (Riquelme et al., 2000). In recent years, increased interest has focused on the search for probiotics that may improve health conditions in the intensive rearing of marine organisms. This research has been spurred by an intention to minimize the use of antibiotics in aquaculture. Apart from the potential hazard of increased antibiotic resistance (Sørum, 1998), the extensive use of antibiotics on fish larvae may result in dramatic changes in the indigenous microflora. In addition to disease prophylaxis, probiotic bacteria may enhance growth rate or increase growth yield of marine larvae by furnishing cell substances or micronutrients such as essential fatty acids, vitamins, minerals, or even enzymes.

7.4 PROBIOTICS IN LARVICULTURE: REARING LIFE

Research on probiotics in larviculture is interesting, promising on the one hand and challenging on the other hand. A number of reviews have been published on the utility and application of probiotics in aquaculture and have suggested that since the only source of exposure to the microbial flora is the rearing environment of the species in aquaculture, the gastrointestinal microflora is quite similar in composition to the nurturing surroundings (Munro et al., 1994; Ringø et al., 1996; Gatesoupe, 1999; Riquelme et al., 2001). A miniscule of knowledge exists in terms of establishment and activity of the microflora, particularly during the larval stages of aquatic organisms (Olafsen, 2001). Using an example of *Artemia* population, it has been proved that the growth depends on both the controllable and uncontrollable (chance) factors, which also determine the composition of resulting gut microflora (Verschuere et al., 1997). The sterile microflora of the larvae (Ringø et al., 1996) attains its inhabitants from either the egg environment at the time of hatching (Hansen and Olafsen, 1990), the water intake for osmoregulation (Reitan et al., 1998; Olafsen, 2001), or when they start feeding (Bergh et al., 1994; Huys et al., 2001).

7.5 PROBIOTICS IN PUBLIC HEALTH: REJUVENATING FITNESS

The rate of increase in resistance to antibiotics (Neu, 1992) is a major public health problem throughout the world (Bengmark, 1998). The immunostimulatory properties of probiotics (Marteau and Rambaud, 1993) offer potential in boosting the host's resistance to disease, thereby potentially reducing the frequency of antibiotic use. The use of probiotics has been proposed by several investigators as an alternative to antibiotics in animals (Underdahl et al., 1982; Fuller, 1989; Canganella et al., 1996; Gardiner et al., 1999) and poultry (Edens et al., 1997; Gardiner et al., 1999). Development of novel marine probiotics for inhibition of pathogens would be a better alternative for this heightened concern of antibiotic resistance.

7.6 MARINE PROBIOTICS: APPROACHES FOR DEVELOPMENT

Probiotics, especially *Lactobacillus* and *Bifidobacterium*, have been suggested to be associated with alleviation of lactose intolerance (Levri et al., 2005); prevention and cure of viral, bacterial, and antibiotic- or radiotherapy-induced diarrheas (Szajewska and Mrukowicz, 2005; Guandalini, 2006; Parvez et al., 2006); immunomodulation (Forsythe and Bienenstock, 2010); antimutagenic (Chalova et al., 2008) and anticarcinogenic effects (Liong, 2008); and even blood cholesterol reduction (Ooi and Liong, 2010). Although probiotics have been employed to treat a number of diseases including certain viral and bacterial diseases, applications of probiotics in other medical arenas are still juvenile. This may be because of lack of research activities on the molecular biology of probiotic strains. Genomic studies on *Lactobacillus* have focused on the interaction with the immune system, anticancer potential, and potential as a biotherapeutic agent in cases of antibiotic-associated diarrhea, traveler's diarrhea, pediatric diarrhea, inflammatory bowel disease, and irritable bowel syndrome. Strategies for the application of biotechnological and molecular biological approaches are required for the development of marine probiotic strains to be used in aquaculture as well as other areas of day to day life.

Another useful probiotic strain, *Bifidobacterium*, can also be explored from the marine environment for the development of the latest and novel probiotic agents. The basic problem faced during the development of marine probiotics is the isolation and identification of the potential strain. Some studies have used 16SrDNA-targeted genus- and species-specific PCR primers for the identification and detection of Bifidobacteria. Identification of cultured Bifidobacteria using PCR primer pairs is rapid and accurate, being based on nucleic acid sequences. Matsuki et al. (2002) have found that, in adult feces, the *Bifidobacterium catenulatum* group is the most commonly detected species, followed by *Bifidobacterium longum*, *Bifidobacterium adolescentis*, and

Bifidobacterium bifidum. In breast-fed infants, *Bifidobacterium breve* was the most frequently detected species, followed by *Bifidobacterium infantis*, *B. longum*, and *B. bifidum* (Matsuki et al., 2002). This knowledge may be useful in specific isolation of these species from the marine environment and exploring their probiotic potential. Real-time, quantitative PCR using primers targeting 16S rDNA has a promising role in the enumeration of *Bifidobacterium*. A similar approach could be applied to the oceanic environment to detect the *Bifidobacterium*, *Lactobacillus* sp., and any other related species of probable probiotic potential. This technique to target bacteria with quantitative PCR will contribute to future studies of the composition and dynamics of the intestinal microflora, thereby giving new insights into the identification of probiotic species to be sought from the marine environment. Identifying a novel marine probiotic strain and providing sufficient numbers of viable bacteria in a product are not the only tasks in this field. Efficient and strong clinical trial should be followed and industry must also provide proof of efficacy for each strain. In the early 1990s, Fonterra embarked on a program to develop proprietary probiotic strains and, as a result, commercialized two strains, *Bifidobacterium lactis* HN019 and *Lactobacillus rhamnosus* HN001 (Crittenden et al., 2005). On similar lines, marine probiotic strains can also be pursued to find a place in the market for consumption as well as utilization in aquaculture. Biomarkers have to be employed to identify strains with probiotic utility and to define the different positive health benefits of existing probiotic strains.

7.7 NANOTECHNOLOGY APPLICATIONS IN AQUACULTURE

Currently, the global concern of post antibiotic era has emerged with reduced capabilities to combat microbial diseases (Blecher et al., 2011). Hence, the development of novel therapeutic approaches to these challenges warrants focal point of modern research (Knipe et al., 2013). The alternative to traditional therapies would be the use of probiotics (Mahasneh and Abbas, 2010; Elbaz and Willner, 2012), quorum sensing regulation, and combination therapy with probiotics and nanoparticles (Day et al., 2009; Taleb et al., 2012). Such methods will definitely improve the outcomes of both traditional and novel bionanotechnology approaches in aquaculture. The fisheries and aquaculture industry can be revolutionized by using nanotechnology with new tools like rapid disease detection, enhancing the ability of fish to absorb drugs like hormones, vaccines, and nutrients rapidly. Interestingly nanotechnology has a wide usage potential in aquaculture and seafood industries (Hardy et al., 2012). This technology can provide novel tools for aquaculture, fish biotechnology, fish genetics, fish reproduction, aquatic health, and detection of fish diseases and produce stronger flavors, color quality, nanoformulations, nanoemulsions, nanoencapsulation, nanovaccines, and nanosensors (Rather et al., 2011).

One recent approach is that nanoparticles will enhance aqua feeds by increasing the proportion of fish food nutrients that pass across the gut tissue and into the fish, rather than passing directly through the fish digestive system unused. Moreover, dietary minerals at the nanoscale level may pass into cells more readily than their larger counterparts and accelerate their process of assimilation into the fish gut. As an example, rainbow trout (*Oncorhynchus mykiss*) exhibited a faster rate of growth when fed with iron nanoparticles (Mohammadi and Tukmechi, 2015). However, researchers at NASA have developed a nanotechnological biosensor that is capable of detecting minute amounts of microbes, including bacteria, viruses, and parasites. This is achieved using highly sensitized carbon nanotubes to detect pathogens in water and food sources (Gregor, 2006).

In this context, the use of nanoparticle carriers like chitosan and poly-lactide-co-glycolide acid of vaccine antigens together with mild inflammatory inducers may give a high level of protection to fishes and shellfishes not only against bacterial diseases but also from certain viral diseases with vaccine-induced side effects (Rajeshkumar et al., 2009).

Nano TiO_2 was demonstrated with exceptional sterilization efficiency on three bacteria of *E. coli, A. hydrophila, and Vibrio anguillarum* (Huang et al., 2010). Interestingly, under ultraviolet irradiation conditions, nano TiO_2 can produce highly active hydroxyl −OH, superoxide ion −O, peroxyl radical −OOH, and other free radicals with high oxidation capacity. These free radicals can interact with biomacromolecules in bacteria, viruses, and other microorganisms and can destroy cell structures through sterilization and disinfection; additionally their efficiency was demonstrated with higher potential compared to traditional bactericides (Zhao et al., 2000; Yu et al., 2002; Sonawane et al., 2003). The harmful and beneficial bacteria in aquaculture were detected by using copper-bearing montmorillonite (Cu^{2+}-MMT) nanotechnologies (Liu et al., 2009). It is interesting to note that bactericidal effects of copper-bearing nano-montmorillonite were also demonstrated with high MIC values on three aquatic pathogenic bacteria and two probiotic bacteria in vitro. Another study demonstrated diet improvement, antioxidant status, glutathione peroxidase activities, and muscle Se concentration in crucian carp (*Carassius auratus gibelio*) when supplemented with (nano-Se and selenomethionine) (Zhou et al., 2009). Another important application of nanotechnologies in seafood safety is by delaying the microbial spoilage by using nanoparticles (Can et al., 2011). Recently, nanoparticles were used as a source of dietary zinc in juvenile grass carp (*Ctenopharyngodon idella*) and compared the growth and hematological indices. Interestingly, this study demonstrated that an efficient drug delivery with functionalized zinc nanoparticles could improve the antiviral effect on grass carp and holds potential application value to control fish viral diseases in aquaculture (Faiz et al., 2015).

7.8 PROMISES AND HOPES

Probiotics have found a number of applications in new functional food products, resulting in the association of more food ingredients with health claims. The marine environment is rich in certain diverse strains of microbes that can turn out to be effective and safe probiotic agents. The existing level of research has expanded the hopes of their use in several areas, including sport-related energy drinks, fortified foods and drinks, and dairy products such as yogurt, cheese, ice cream, and milk containing prebiotics and probiotics. The market for probiotics is blooming, but its sustainability greatly depends upon further research, scientific substantiation, and biotechnological intervention in exploring the marine environment. Some probiotic strains were shown to inhibit the growth of enteropathogens, such as *Salmonella enteritidis*, enterotoxigenic *E. coli*, and *Serratia marcescens*, in vitro (Gonzalez et al., 1993; Drago et al., 1997) and in this regard may offer considerable therapeutic potential. Another report suggests that *Lactobacillus* exerts antagonist activity against *Salmonella typhimurium* C5 infection both in vitro and in vivo (Hudault et al., 1997). These studies provide evidence and form a basis for the clinical use of probiotics in suppression of pathogens. Apart from the aforementioned antimicrobial activity of therapeutic importance, potential clinical applications for probiotic microorganisms include treatment of food allergy (Salminen et al., 1996), reduction of hypertension (Hata et al., 1996), and usage as vectors for delivery of oral vaccines

(Pouwels et al., 1996). An effective regime for studies related to food-grade genetic markers (Allison and Klaenhammer, 1996; Lin et al., 1996) would be quite beneficial for the construction of probiotics against antibiotic-resistant genes.

Probiotics not just maintain the health of living creatures but may also protect the urogenital tract against microbial infections (Reid et al., 1994, 1998; Reid and Bruce, 2006). Certain probiotic strains are also shown to reduce the risk of infections associated with the use of medical devices. Another potential application of probiotic cultures is in the production of fermented food products enriched in health-promoting substances, such as conjugated linoleic acid (CLA). The beneficial effects of CLA include anticarcinogenic activity, antiatherogenic activity, the ability to reduce the catabolic effects of immune stimulation, the ability to enhance growth promotion, and the ability to reduce body fat. Foods that contain animal fat, such as beef, lamb, milk, and dairy foods, are rich sources of CLA (Stanton et al., 1997). Certain strains of *Propionibacteria*, commonly used as dairy starter cultures, were shown to be able to convert free linoleic acid to CLA (Jiang et al., 1998), suggesting that it may be possible via fermentation to produce fermented food products enriched in CLA. Probiotics may hold a number of other potential applications that are still uncovered due to lack in technological interventions in this field.

7.9 FUTURE DIRECTIONS: RAY OF HOPE

It is becoming increasingly accepted by consumers that live lactic acid bacteria do exert health benefits when eaten. In fact, many countries consume live lactic acid bacteria in the form of traditional food termed as "curd." Further, it is strongly believed that all probiotic bacteria do not act in the same fashion; neither do their activities have the same magnitude. There is an imperative need to search for newer, safer, and effective probiotics with a broad range of health benefits. The marine environment holds a plethora of bioactive compounds with different bioactivities. The marine microbes have also been reported to hold great promise as antimicrobial agents and antitumor agents (Bhatnagar and Kim, 2010a,b). It is not hard to develop methods for effective isolation and characterization of some of these marine strains to be promising probiotic strains. Although biomarker research can prove to be a very efficient tool, which biomarkers accurately reflect efficacy in vivo is in question as the mechanisms by which most probiotic bacteria exert their health benefits remain unclear (Dekker et al., 2007). Recent technological advances have turned their wheels to unravel how probiotic bacteria work. Once we are able to define a set of mechanism of action to probiotic strains, biomarker development would be easier with a promise to offer the market safe and effective probiotic medicinal foods.

REFERENCES

Allison, G. and Klaenhammer, T. Functional analysis of the gene encoding immunity to lactacin F, lafI, and its use as a *Lactobacillus*-specific, food-grade genetic marker. *Appl. Environ. Microbiol.*, 62 (1996): 4450–4460.

Austin, B., Stuckey, L., Robertson, P., Effendi, I., and Griffith, D. A probiotic strain of *Vibrio alginolyticus* effective in reducing diseases caused by *Aeromonas salmonicida, Vibrio anguillarum* and *Vibrio ordalii. J. Fish Dis.*, 18 (2006): 93–96.

Bengmark, S. Ecological control of the gastrointestinal tract. The role of probiotic flora. *Gut*, 42 (1998): 2.

Bergh, Ø., Naas, K.E., and Harboe, T. Shift in the intestinal microflora of Atlantic halibut (*Hippoglossus hippoglossus*) larvae during first feeding. *Can. J Fish. Aquat. Sci.*, 51 (1994): 1899–1903.

Bhatnagar, I. and Kim, S.K. Immense essence of excellence: Marine microbial bioactive compounds. *Mar. Drugs*, 8 (2010a): 2673–2701.

Bhatnagar, I. and Kim, S.K. Marine antitumor drugs: Status, shortfalls and strategies. *Mar. Drugs*, 8 (2010b): 2702–2720.

Blecher, K., Nasir, A., and Friedman, A. The growing role of nanotechnology in combating infectious diseases. *Virulence*, 2(5) (2011): 395–401.

Bondad-Reantaso, M.G., Subasinghe, R.P., Arthur, J.R., Ogawa, K., Chinabut, S., Adlard, R., Tan, Z., and Shariff, M. Disease and health management in Asian aquaculture. *Vet. Parasitol.*, 132 (2005): 249–272.

Can, E., Kizak, V., Kayim, M., Can, S.S., Kutlu, B., Ates, M., Kocabas, M., and Demirtas, N. Nanotechnological applications in aquaculture-seafood industries and adverse effects of nanoparticles on environment. *J. Mater. Sci. Eng.*, 5(5) (2011).

Canganella, F., Gasbarri, M., Massa, S., and Trovatelli, L. A microbiological investigation on probiotic preparations used for animal feeding. *Microbiol. Res.* 151 (1996): 167–175.

Chalova, V., Lingbeck, J., Kwon, Y., and Ricke, S. Extracellular antimutagenic activities of selected probiotic *Bifidobacterium* and *Lactobacillus* spp. as a function of growth phase. *J. Environ. Sci. Health Part B*, 43 (2008): 193–198.

Crittenden, R., Bird, A., Gopal, P., Henriksson, A., Lee, Y., and Playne, M. Probiotic research in Australia, New Zealand and the Asia-Pacific region. *Curr. Pharm. Des.*, 11 (2005): 37–53.

Day, E.S., Morton, J.G., and West, J.L. Nanoparticles for thermal cancer therapy. *J. Biomech. Eng.*, 131(7) (2009): 074001.

Dekker, J., Collett, M., Prasad, J., and Gopal, P. Functionality of probiotics—Potential for product development. *Forum Nutr.*, 60 (2007): 196.

Drago, L., Gismondo, M.R., Lombardi, A., Haën, C., and Gozzini, L. Inhibition of in vitro growth of enteropathogens by new *Lactobacillus* isolates of human intestinal origin. *FEMS Microbiol. Lett.*, 153 (1997): 455–463.

Driks, A. *Bacillus subtilis* spore coat. *Microbiol. Mol. Boil. Rev.*, 63 (1999): 1–20.

Edens, F., Parkhurst, C., Casas, I., and Dobrogosz, W. Principles of ex ovo competitive exclusion and in ovo administration of *Lactobacillus reuteri*. *Poult. Sci.*, 76 (1997): 179.

Elbaz, J. and Willner, I. DNA origami: Nanorobots grab cellular control. *Nat. Mater.*, 11 (2012): 276–277.

Faiz, H., Zuberi, A., Nazir, S., Rauf, M., and Younus, N. Zinc oxide, zinc sulfate and zinc oxide nanoparticles as source of dietary zinc: Comparative effects on growth and hematological indices of juvenile grass carp (*Ctenopharyngodon idella*). *Int. J. Agric. Biol.*, 17(3) (2015).

FAO Fisheries and Aquaculture Department. The state of world fisheries and aquaculture. *Food and Agriculture Organization of the United Nations*, Rome, (2009), 196 p.

Fonseca, F., Béal, C., and Corrieu, G. Operating conditions that affect the resistance of lactic acid bacteria to freezing and frozen storage. *Cryobiology*, 43 (2001): 189–198.

Forsythe, P. and Bienenstock, J. Immunomodulation by commensal and probiotic bacteria. *Immunol. Invest.*, 39 (2010): 429–448.

Fuller, R. Probiotics in man and animals. *J. Appl. Bacteriol.*, 66 (1989): 365.

Gardiner, G., Stanton, C., Lynch, P., Collins, J., Fitzgerald, G., and Ross, R. Evaluation of cheddar cheese as a food carrier for delivery of a probiotic strain to the gastrointestinal tract. *J. Dairy Sci.*, 82 (1999): 1379–1387.

Gatesoupe, F. The use of probiotics in aquaculture. *Aquaculture*, 180 (1999): 147–165.

Gatesoupe, F.J. The effect of three strains of lactic bacteria on the production rate of rotifers, *Brachionus plicatilis*, and their dietary value for larval turbot, *Scophthalmus maximus*. *Aquaculture*, 96 (1991): 335–342.

Gatesoupe, F.J. Lactic acid bacteria increase the resistance of turbot larvae, *Scquhthalmus maximus*, against pathogenic *Vibrio*. *Aquat. Living Resour.*, 7 (1994): 277–282.

Gatesoupe, F.J. Updating the importance of lactic acid bacteria in fish farming: Natural occurrence and probiotic treatments. *J. Mol. Microbiol. Biotechnol.*, 14 (2008): 107–114.

Gibson, L., Woodworth, J., and George, A. Probiotic activity of *Aeromonas media* on the Pacific oyster, *Crassostrea gigas*, when challenged with *Vibrio tubiashii*. *Aquaculture*, 169 (1998): 111–120.

Gildberg, A. and Mikkelsen, H. Effects of supplementing the feed to Atlantic cod (*Gadus morhua*) fry with lactic acid bacteria and immuno-stimulating peptides during a challenge trial with *Vibrio anguillarum*. *Aquaculture*, 167 (1998): 103–113.

Gismondo, M.R., Drago, L., and Lombardi, A. Review of probiotics available to modify gastrointestinal flora. *Int. J. Antimicrob. Agents*, 12 (1999): 287–292.

Gomez-Gil, B., Roque, A., and Turnbull, J.F. The use and selection of probiotic bacteria for use in the culture of larval aquatic organisms. *Aquaculture*, 191 (2000): 259–270.

Gonzalez, S., Apella, M., Romero, N., De Macias, M.E.N., and Oliver, G. Inhibition of enteropathogens by *Lactobacilli* strains used in fermented milk. *J. Food Protect.*, 56 (1993): 773–776.

Gregor, S. The Nature of Theory in Information Systems. *MIS Quarterly*, 30(3) (2006): 611–642.

Guandalini, S. Probiotics for children: Use in diarrhea. *J. Clin. Gastroenterol.*, 40 (2006): 244–248.

Gullian, M., Thompson, F., and Rodriguez, J. Selection of probiotic bacteria and study of their immunostimulatory effect in *Penaeus vannamei*. *Aquaculture*, 233 (2004): 1–14.

Hamilton-Miller, J., Shah, S., and Winkler, J. Public health issues arising from microbiological and labelling quality of foods and supplements containing probiotic micro-organisms. *Public Health Nutr.*, 2 (1999): 223–229.

Hansen, G.H. and Olafsen, J.A. Bacterial colonization of cod (*Gadus morhua* L.) and halibut (*Hippoglossus hippoglossus*) eggs in marine aquaculture. *Appl. Environ. Microbiol.*, 55 (1989): 1435–1446.

Hansen, G.H. and Olafsen, J.A. Endocytosis of bacteria in yolksac larvae of cod (*Gadus morhua* L.). *Microbiol. Poecilotherms*, (1990): 187–191.

Hardy, M.C. Using selective insecticides in sustainable IPM, *Plant Sciences Reviews* 2011 (2012): 127.

Hata, Y., Yamamoto, M., Ohni, M., Nakajima, K., Nakamura, Y., and Takano, T. A placebo-controlled study of the effect of sour milk on blood pressure in hypertensive subjects. *Am. J. Clin. Nutr.*, 64 (1996): 767.

Hjelm, M., Bergh, Ø., Riaza, A., Nielsen, J., Melchiorsen, J., Jensen, S., Duncan, H., Ahrens, P., Birkbeck, H., and Gram, L. Selection and identification of autochthonous potential probiotic bacteria from turbot larvae (*Scophthalmus maximus*) rearing units. *Syst. Appl. Microbiol.*, 27 (2004): 360–371.

Holmström, K., Gräslund, S., Wahlström, A., Poungshompoo, S., Bengtsson, B.E., and Kautsky, N. Antibiotic use in shrimp farming and implications for environmental impacts and human health. *Int. J. Food Sci. Technol.*, 38 (2003): 255–266.

Huang, F., Chen, D., Zhang, X.L., Caruso, R.A., Cheng, Y.B. Dual-function scattering layer of submicrometer-sized mesoporous TiO2 Beads for high-efficiency dye-sensitized solar cells. *Adv. Funct. Mater.*, 20(8) (2010): 1301–1305.

Hudault, S., Lievin, V., Bernet-Camard, M.F., and Servin, A.L. Antagonistic activity exerted in vitro and in vivo by *Lactobacillus casei* (strain GG) against *Salmonella typhimurium* C5 infection. *Appl. Environ. Microbiol.*, 63 (1997): 513.

Hughes, V.L. and Hillier, S.L. Microbiologic characteristics of *Lactobacillus* products used for colonization of the vagina. *Obstetr. Gynecol.*, 75 (1990): 244.

Huys, L., Dhert, P., Robles, R., Ollevier, F., Sorgeloos, P., and Swings, J. Search for beneficial bacterial strains for turbot (*Scophthalmus maximus* L.) larviculture. *Aquaculture*, 193 (2001): 25–37.

Irianto, A. and Austin, B. Use of probiotics to control furunculosis in rainbow trout, *Oncorhynchus mykiss* (Walbaum). *J. Fish Dis.*, 25 (2002): 333–342.

Jiang, J., Björck, L., and Fonden, R. Production of conjugated linoleic acid by dairy starter cultures. *J. Appl. Microbiol.*, 85(1) (1998): 95–102.

Karunasagar, I., Pai, R., and Malathi, G. Mass mortality of *Penaeus monodon* larvae due to antibiotic-resistant *Vibrio harveyi* infection. *Aquaculture*, 128 (1994): 203–209.

Kesarcodi-Watson, A., Kaspar, H., Lategan, M.J., and Gibson, L. Probiotics in aquaculture: The need, principles and mechanisms of action and screening processes. *Aquaculture*, 274 (2008): 1–14.

Knipe, J.M., Peters, J.T., and Peppas, N.A. Theranostic agents for intracellular gene delivery with spatiotemporal imaging. *NanoToday*, 8(1) (2013): 21–38.

Kozasa, M. Toyocerin (*Bacillus toyoi*) as growth promotor for animal feeding. *Microbiol. Aliment. Nutr.*, 4 (1986): 121–135.

Lategan, M. and Gibson, L. Antagonistic activity of *Aeromonas media* strain A199 against *Saprolegnia* sp., an opportunistic pathogen of the eel, *Anguilla australis* Richardson. *J. Fish Dis.*, 26 (2003): 147–153.

Levri, K.M., Ketvertis, K., Deramo, M., Merenstein, J.H., and D Amico, F. Do probiotics reduce adult lactose intolerance? A systematic review. *J. Fam. Pract.*, 54 (2005): 613.

Lin, M.Y., Harlander, S., and Savaiano, D. Construction of an integrative food-grade cloning vector for *Lactobacillus acidophilus*. *Appl. Microbiol. Biotechnol.*, 45 (1996): 484–489.

Liong, M.T. Roles of probiotics and prebiotics in colon cancer prevention: Postulated mechanisms and in-vivo evidence. *Int. J. Mol. Sci.*, 9 (2008): 854–863.

Liu, P.W., Guo, T., and Wei, H. Bactericidal effects of copper-bearing nano-montmorillonite on three aquatic pathogenic bacteria and two intestinal available bacteria in vitro. *J. Shanghai Ocean Univ.*, 5 (2009): 520–526. Article number: 1674-5566(2009)05-0520-07.

Mahasneh, A.M. and Abbas, M.M. Probiotics and traditional fermented foods: The eternal connection. *Jordan J. Biol. Sci.*, 3(4) (2010): 133–140.

Marteau, P. and Rambaud, J.C. Potential of using lactic acid bacteria for therapy and immunomodulation in man. *FEMS Microbiol. Rev.*, 12 (1993): 207–220.

Matsuki, T., Watanabe, K., and Tanaka, R. Genus-and species-specific PCR primers for the detection and identification of *Bifidobacteria*. In: *Probiotics and Prebiotics: Where Are We Going?* Tannock, G.W., Ed. Caister Academic Press, Wymondham, UK, 2002, p. 85.

Metchnikoff, E. Lactic acid as inhibiting intestinal putrefaction. In: *The Prolongation of Life: Optimistic Studies*. Chalmers Mitchell, P., Ed. Heinemann, London, UK, 1907, pp. 161–183.

Mohammadi, N. and Tukmechi, A. The effects of iron nanoparticles in combination with *Lactobacillus casei* on growth parameters and probiotic counts in rainbow trout (*Oncorhynchus mykiss*) intestine. *J. Vet. Res.*, 70(1) (2015): Pe47–Pe53.

Mohanty, S., Swain, S., and Tripathi, S. Growth and survival of rohu spawn fed on a liver based diet. *J. Inland Fish. Soc. India*, 25 (1993): 41–45.

Mohanty, S., Swain, S., and Tripathi, S. Rearing of catla (*Catla catla Ham.*) spawn on formulated diets. *J. Aquacul. Tropics*, 11 (1996): 253–258.

Moriarty, D. Control of luminous *Vibrio* species in penaeid aquaculture ponds. *Aquaculture*, 164 (1998): 351–358.

Munro, P., Barbour, A., and Blrkbeck, T. Comparison of the gut bacterial flora of start-feeding larval turbot reared under different conditions. *J. Appl. Microbiol.*, 77 (1994): 560–566.

Neu, H.C. The crisis in antibiotic resistance. *Science*, 257 (1992): 1064.

Nomoto, K. Prevention of infections by probiotics. *J. Biosci. Bioeng.*, 100 (2005): 583–592.

Oggioni, M.R., Ciabattini, A., Cuppone, A.M., and Pozzi, G. *Bacillus* spores for vaccine delivery. *Vaccine*, 21 (2003): S96–S101.

Olafsen, J.A. Interactions between fish larvae and bacteria in marine aquaculture. *Aquaculture*, 200 (2001): 223–247.

Ooi, L.G. and Liong, M.T. Cholesterol-lowering effects of probiotics and prebiotics: A review of in vivo and in vitro findings. *Int. J. Mol. Sci.*, 11 (2010): 2499–2522.

Parker, R.B. Probiotics, the other half of the antibiotics story. *Anim. Nutr. Health*, 29 (1974): 4–8.

Parvez, S., Malik, K., Ah Kang, S., and Kim, H.Y. Probiotics and their fermented food products are beneficial for health. *J. Appl. Microbiol.*, 100 (2006): 1171–1185.

Phianphak, W., Rengpipat, S., Piyatiratitivorakul, S., and Menasveta, P. Probiotic use of *Lactobacillus* spp. for black tiger shrimp, *Penaeus monodon*. *J. Sci. Res. Chula. Univ.*, 24 (1999): 41–58.

Pouwels, P.H., Leer, R.J., and Boersma, W.J.A. The potential of *Lactobacillus* as a carrier for oral immunization: Development and preliminary characterization of vector systems for targeted delivery of antigens. *J. Biotechnol.*, 44 (1996): 183–192.

Queiroz, J.F. and Boyd, C.E. Effects of a bacterial inoculum in channel catfish ponds. *J. World Aquacul. Soc.*, 29 (2007): 67–73.

Rajeshkumar, S., Venkatesan, C., Sarathi, M., Sarathbabu, V., Thomas, J., Anver Basha, K., and Sahul Hameed, A.S. Oral delivery of DNA construct using chitosan nanoparticles to protect the shrimp from white spot syndrome virus (WSSV). *Fish Shellfish Immunol.*, 26 (2009): 429–437.

Rather, M., Sharma, R., Aklakur, M., Ahmad, S., Kumar, N., Khan, M., and Ramya, V.L. Nanotechnology: A novel tool for aquaculture and fisheries development. *Fish. Aquacul. J.*, 2011 (2011): 16.

Reid, G. The importance of guidelines in the development and application of probiotics. *Curr. Pharm. Des.*, 11 (2005): 11–16.

Reid, G. and Bruce, A.W. Probiotics to prevent urinary tract infections: The rationale and evidence. *World J. Urol.*, 24 (2006): 28–32.

Reid, G., Bruce, A.W., and Smeianov, V. The role of *Lactobacilli* in preventing urogenital and intestinal infections. *Int. Dairy J.*, 8 (1998): 555–562.

Reid, G., Lam, D., Bruce, A.W., van der Mei, H.C., and Busscher, H.J. Adhesion of *Lactobacilli* to urinary catheters and diapers: Effect of surface properties. *J. Biomed. Mater. Res.*, 28 (1994): 731–734.

Reitan, K., Natvik, C., and Vadstein, O. Drinking rate, uptake of bacteria and microalgae in turbot larvae. *J. Fish Biol.*, 53 (1998): 1145–1154.

Rengpipat, S., Phianphak, W., Piyatiratitivorakul, S., and Menasveta, P. Effects of a probiotic bacterium on black tiger shrimp *Penaeus monodon* survival and growth. *Aquaculture*, 167 (1998): 301–313.

Ringø, E., Birkbeck, T., Munro, P., Vadstein, O., and Hjelmeland, K. The effect of early exposure to *Vibrio pelagius* on the aerobic bacterial flora of turbot, *Scophthalmus maximus* (L.) larvae. *J. Appl. Microbiol.*, 81 (1996): 207–211.

Riquelme, C., Araya, R., and Escribano, R. Selective incorporation of bacteria by *Argopecten purpuratus* larvae: Implications for the use of probiotics in culturing systems of the *Chilean scallop*. *Aquaculture*, 181 (2000): 25–36.

Riquelme, C.E., Jorquera, M.A., Rojas, A.I., Avendaño, R.E., and Reyes, N. Addition of inhibitor-producing bacteria to mass cultures of *Argopecten purpuratus* larvae (Lamarck, 1819). *Aquaculture*, 192 (2001): 111–119.

Robertson, P., O'Dowd, C., Burrells, C., Williams, P., and Austin, B. Use of *Carnobacterium* sp. as a probiotic for Atlantic salmon (*Salmo salar* L.) and rainbow trout (*Oncorhynchus mykiss*, Walbaum). *Aquaculture*, 185 (2000): 235–243.

Ruiz-Ponte, C., Samain, J., Sanchez, J., and Nicolas, J. The benefit of a *Roseobacter* species on the survival of scallop larvae. *Mar. Biotechnol.*, 1 (1999): 52–59.

Salminen, S., Isolauri, E., and Salminen, E. Probiotics and stabilisation of the gut mucosal barrier. *Asia Pac. J. Clin. Nutr.*, 5 (1996): 53–56.

Schillinger, U. Isolation and identification of *Lactobacilli* from novel-type probiotic and mild yoghurts and their stability during refrigerated storage. *Int. J. Food Microbiol.*, 47 (1999): 79–87.

Schulze, A.D., Alabi, A.O., Tattersall-Sheldrake, A.R., and Miller, K.M. Bacterial diversity in a marine hatchery: Balance between pathogenic and potentially probiotic bacterial strains. *Aquaculture*, 256 (2006): 50–73.

Sharma, O. and Bhukar, S. Effect of Aquazyn-TM-1000, a probiotic on the water quality and growth of *Cyprinus carpio var. communis* (L.). *Indian J. Fish.*, 47 (2011): 209–213.

Skjermo, J. and Vadstein, O. Techniques for microbial control in the intensive rearing of marine larvae. *Aquaculture*, 177 (1999): 333–343.

Sonawane, R.S., Hegde, S.G., and Dongare, M.K. Preparation of titanium (VI) oxide thin film photocatalyst by sol–gel dip coating. *Mater. Chem. Phys.*, 3 (2003): 744–746.

Sørum, H. Mobile drug resistance genes among fish bacteria. *Acta Pathologica, Microbiologica et Immunologica Scandinavica (APMIS)*, 106(84) (1998): 74–76.

Stanton, C., Lawless, F., Murphy, J., and Connolly, J. Conjugated linoleic acid—A marketing advantage for animal fats. In: *Animal Fats: BSE and After*. Berger, K.G., Ed. Barnes and Assoc, Bridgwater, UK, 1997, pp. 27–41.

Szajewska, H. and Mrukowicz, J.Z. Use of probiotics in children with acute diarrhea. *Pediatr. Drugs*, 7 (2005): 111–122.

Talib, W.H., AbuZarga, M.H., and Mahasneh, A.M. Antiproliferative, antimicrobial and apoptosis inducing effects of compounds isolated from *Inula viscosa*. *Molecules*, 17(3) (2012): 3291–3303.

Underdahl, N., Torres-Medina, A., and Dosten, A. Effect of *Streptococcus faecium* C-68 in control of *Escherichia coli*-induced diarrhea in gnotobiotic pigs. *Am. J. Vet. Res.*, 43 (1982): 2227.

Vaseeharan, B. and Ramasamy, P. Control of pathogenic *Vibrio* spp. by *Bacillus subtilis* BT23, a possible probiotic treatment for black tiger shrimp *Penaeus monodon. Lett. Appl. Microbiol.*, 36 (2003): 83–87.

Verschuere, L., Dhont, J., Sorgeloos, P., and Verstraete, W. Monitoring biolog patterns and r/K-strategists in the intensive culture of Artemia juveniles. *J. Appl. Microbiol.*, 83 (1997): 603–612.

Verschuere, L., Rombaut, G., Sorgeloos, P., and Verstraete, W. Probiotic bacteria as biological control agents in aquaculture. *Microbiol. Mol. Biol. Rev.*, 64 (2000): 655–671.

Vieira, F.N., Buglione, C.C., Mouriño, J.P.L., Jatobá, A., Martins, M.L., Schleder, D.D., Andreatta, E.R., Barraco, M.A., and Vinatea, L.A. Effect of probiotic supplemented diet on marine shrimp survival after challenge with *Vibrio harveyi. Arq. Bras. Med. Vet. Zoot.*, 62 (2010): 631–638.

Vine, N.G., Leukes, W.D., and Kaiser, H. Probiotics in marine larviculture. *FEMS Microbiol. Rev.*, 30 (2006): 404–427.

Wang, Y.B. Effect of probiotics on growth performance and digestive enzyme activity of the shrimp *Penaeus vannamei. Aquaculture*, 269 (2007): 259–264.

Wang, Y.B. and Xu, Z.R. Probiotics treatment as method of biocontrol in aquaculture. *Feed Res.*, 12 (2004): 42–45.

Wang, Y.B., Xu, Z.R., and Xia, M.S. The effectiveness of commercial probiotics in northern white shrimp *Penaeus vannamei* ponds. *Fish. Sci.*, 71 (2005): 1036–1041.

Yanbo, W. and Zirong, X. Effect of probiotics for common carp (*Cyprinus carpio*) based on growth performance and digestive enzyme activities. *Anim. Feed Sci. Technol.*, 127 (2006): 283–292.

Yu, J.C., Tang, H.Y., and Yu, J.G. Bactericidal and photocatalytic activities of TiO2 thin films prepared by sol–gel and reverse micelle methods. *J. Photochem. Photobiol. A Chem.*, 3 (2002): 211–219.

Zhao, D., Wang, J., Sun, B.H., Sun, B.H., Gao, J.Q., and Xu, R. Development and application of TiO2 photocatalysis as antimicrobial agent. *J. Liaoning Univ. (Nat. Sci. Ed.)*, 2 (2000): 173–174.

Zhong, W., Millsap, K., Bialkowska-Hobrzanska, H., and Reid, G. Differentiation of *Lactobacillus* species by molecular typing. *Appl. Environ. Microbiol.*, 64 (1998): 2418.

Zhou, X., Wang, Y., Gu, Q., and Li, W. Effects of different dietary selenium sources (selenium nanoparticle and selenomethionine) on growth performance, muscle composition and glutathione peroxidase enzyme activity of crucian carp (*Carassius auratus gibelio*). *Aquaculture*, 291 (2009): 78–81.

Index

A

Actinobacteria
  actinomycete antitumor action, 9
  algae, 6–7
  antibacterial action, 9–10
  antidiabetic activity, 11
  antifungal compounds, 10
  anti-HIV activity, 12
  anti-inflammatory activity, 11
  antimalarial activity, 11
  antiviral compounds, 10
  cytostatic activity, 11
  cytotoxic activity, 10
  seaweeds, 7–8
  secondary metabolites, 7
  sponges, 8–9
Actinomycete antitumor action, 9
Agar well diffusion technique, 28, 83
Aldoses, 38
Algae, 6–8, 10–12, 26, 29, 36, 44, 77–79, 100, 108, 113, 115, 117–119
α-methyl glycoside, 38
Ammonium sulfate precipitation, 84–86, 119, 121
Amphotericin B, 10
Anion exchange chromatography, 40, 43, 88
Anionic exchangers, 87–88
Antibacterial action, 9–10
Antibiotics, 4, 6, 68, 79–80, 82, 130, 135–136
Antidiabetic activity, 11
Antifungal compounds, 10
Anti-HIV activity, 12
Antimicrobial Peptide Database, 92
Antiviral compounds, 10
Aphanizomenon flos-aquae, 102
Astaxanthin, 100–102, 104, 108, 112–118
Attached EPS, 37

B

Bacillus spp., 5, 78, 80–81, 133–135
Bacteria, 5–6, 20–29, 37, 39–40, 46, 78–80, 82, 92, 129–135, 137–139
Bacterial-exopolymeric substances (EPS), 37
Bacteriocins, 5, 80, 82, 130
Bactibase, 92
Bagel, 92

C

β-carotene, 100–102, 112, 114–118
Bergys Manual of Determinative Bacteriology, 22
Bifidobacterium sp., 136–137, 139
Bioactive peptides, *see* Peptides
Bioactive proteins, 81–82
Biodiesel, 61–63, 66, 68, 70
Biomass
  lipid extraction
    dry, 63, 66
    microalgae cell disruption, 66–69
    organic solvents, 63–65
    wet, 66
Biomedical bioactive compounds
  antibacterial activity, 27–28
  anticancer activity, 28–29
  antifungal activity, 28
  anti-HIV activity, 28
  antioxidant activity, 29
Biosurfactants, 5, 25, 79
β-methyl glycoside, 38
Botryococcus sp., 65, 68–69, 108

Calicheamicins, 9
Candida albicans, 10
Capillary gas chromatography (CGC), 46
Capsular EPS, 37
Carbohydrates; *see also* Starch
  aldoses, 38
  analysis, 39–40
  composition, 39
  detection
    furfural, transformation to, 40–41
    5-hydroxy-1-tetralone, 41
    stannous chloride, sulfuric acid, and urea, 42
    triphenyltetrazolium chloride, 41
  disaccharides, 38
  function, 39
  glycosides, 38
  hexosans, 39
  ketoses, 38
  mutarotation, 38
  neutral sugars, 38
  pentosans, 39

reactive sites, 40
simple sugars/monosaccharides, 37
Carnobacterium, 133–134
Carotenes, 101–102
Carotenoids, 8, 63
  classical solvent extraction, 108, 112–113
  microalgal pigments, 101–102
  PFE, 113–115
  supercritical fluid extraction, 115–118
Cationic exchangers, 87
Cavitation, 45, 68, 113
Cell disruption
  lipid extraction
    bead milling, 67
    dry biomass, 63, 66
    effect, 67–69
    high-pressure homogenization, 68
    methods, 66–67
    microalgae, 66–67
    nonmechanical methods, 68
    ultrasonication process, 68
    wet biomass, 66
Cellulose, 24, 39, 41, 47, 88, 108
CGC, *see* Capillary gas chromatography
Chlorella ellipsoidea, 113–114
Chlorella vulgaris, 63–64, 68–69, 103, 114
Chlorococcum sp., 64, 69, 108, 112
Chlorophyll
  microalgal pigments, 100–101
    chlorophyll extraction, 104–105
    chromatographic techniques, 108–111
    classical solvent extraction, 105–106
    SFE, 107
cis-β-carotene, 117–118
CLA, *see* Conjugated linoleic acid
Colorimetric/spectrophotometric
    method, 46–47
Conjugated linoleic acid (CLA), 139
C-phycocyanin (CPC), 102, 118–121
Crustaceans, 37, 134–135
Curd, 139
Cyclomarin A, 11
Cylindrotheca closterium, 104

D
Dialysis/diafiltration, 42–43
Disaccharides, 38–39, 47
Dissolved organic matter (DOM), 48–49
DNS derivatization, 49
Dry biomass, 63, 66
Dubois method, *see* Phenol-sulfuric acid (PSA)
    method
Dunaliella tertiolecta, 104

E
Electrospray ionization (ESI)-MS, 91
Enhanced solvent extraction, *see* Pressurized
    liquid extraction
Enzymatic techniques, 50
Enzymes, 23–24, 46, 50, 61, 63, 67–68,
    79–80, 108
  digestion, 43, 134–135
  inhibitors, 24
Exopolymeric substances, 37
Exopolysaccharides (EPS), 5, 25
Extracellular polymeric substances
  isolation and purification
    anion exchange chromatography, 43
    dialysis/diafiltration, 42–43
    enzyme digestion, 43
    MAE, 45–46
    PHWE, 45
    PLE, 45
    precipitation, 43
    SDS-PAGE, 44
    SFE, 44
    single-dimension gel electrophoresis,
        43–44
    size exclusion chromatography, 43
    UAE, 45
Extremophiles isolation, 21–22

F
Fast atom bombardment (FAB)-MS, 91
Fish
  eggs and larvae, 133–134
  juveniles and adults, 134
Fructose, 38, 41–42, 48
Fungi, 5–6, 20, 22–23, 27–29, 36, 39,
    78, 80

G
Gas chromatography-flame ionization
    detection (GC-FID), 40, 49–50
Gas chromatography-mass spectrometry
    (GC-MS), 26, 46
GC-FID, *see* Gas chromatography-flame
    ionization detection
Gel permeation, 88–89
Glycine-SDS-PAGE, 90–91
Glycosides, 38

H
Haematococcus pluvialis, 108, 113–115,
    117–118
Halophiles, 21–22
Hexosans, 39

High-pressure anion exchange chromatography (HPAEC), 48–49
High-pressure liquid chromatography (HPLC), 46, 48, 50, 108–110
High-pressure solvent extraction, *see* Pressurized liquid extraction (PLE)
High-speed homogenization, 68
HIV/AIDS, *see* Human immunodeficiency infection
HPAEC, *see* High-pressure anion exchange chromatography
HPLC, *see* High-pressure liquid chromatography
Human immunodeficiency virus (HIV), 12, 28
Hydrolytic enzymes, 23
Hydrophobic resins, 86

I
IEF, *see* Isoelectric focusing electrophoresis
Industrial bioactive compounds
  biosurfactants, 25
  enzyme inhibitors, 24
  enzymes, 23–24
  exopolysaccharides, 25
  marine probiotics, 26–27
  pigments, 26
  PUFA, 25–26
Invert sugar, 38
Ion exchange chromatography, 86–88
Isoelectric focusing electrophoresis (IEF), 51
Isolation
  coral reefs, 21
  extremophiles, 21–22
  from mangrove ecosystems, 21
  media, 22–23
  procedures, 20

K
Ketoses, 38
Kirby–Bauer well diffusion method, *see* Agar well diffusion technique

L
Lactobacillus, 133–134, 136–138
Larviculture, 135–136
Leptolyngbya sp. KC45, 119
Lipid extraction
  dry biomass, 63, 66
  microalgae cell disruption
    bead milling, 67

high-pressure homogenization, 68
  methods, 66–67
  nonmechanical methods, 68
  ultrasonication process, 68
nanotechnology, 70
organic solvents, 63–65
wet biomass, 66
Live lactic acid bacteria, 139
Lutein, 100–104, 112, 114–118

M
MAE, *see* Microwave-assisted extraction
MALDI, 92
Marine actinobacteria
  actinomycete antitumor action, 9
  algae, 6–7
  antibacterial action, 9–10
  antidiabetic activity, 11
  antifungal compounds, 10
  anti-HIV activity, 12
  anti-inflammatory activity, 11
  antimalarial activity, 11
  antiviral compounds, 10
  cytostatic activity, 11
  cytotoxic activity, 10
  seaweeds, 7–8
  secondary metabolites, 7
  sponges, 8–9
Marine bacteria, 5–6
Marine environment, 76–77
Marine fungi, 5–6
Marine microbiota, 78
Mascot algorithm (Matrix Science), 92
Mass spectrometry, 50, 91
Matrix/prepacked columns, peptides, 88, 90
MBTH (3-methyl-2-benzothiazolinone hydrazone hydrochloride), 40, 47
Methanopyrus kandleri, 21
Methyl esters, 61–62
Microalgae, 78–79
  biodiversity, 100
  cell disruption methods
    MAE, 104
    PEF for pigment extraction, 103–104
  extraction and purification
    carotenoid, 108–118
    chlorophyll, 104–108
    phycobiliproteins, 118–122
  extraction techniques, 100
  pigments, 100–101
    carotenoids, 101–102
    chlorophyll, 101
    phycobiliproteins, 102

Microalgae cell disruption
   lipid extraction
      bead milling, 67
      dry biomass, 63, 66
      high-pressure homogenization, 68
      methods, 66–67
      nonmechanical methods, 68
      ultrasonication process, 68
      wet biomass, 66
Microbial media, 22–23
Micromonospora echinospora, 9
Microwave-assisted extraction (MAE), 45–46,
   104
Mollusks, 6, 77, 135
Mutarotation, 38

N
Nannochloropsis gaditana, 107
Nanotechnology, aquaculture, 137–138
Neutral carbohydrates/monosaccharides, *see*
   Neutral sugars
Neutral sugars, 38
Nonattached EPS, 37
Nutraceuticals
   bioactive peptides, 81–82
   definition, 76
   marine environment, 77
   marine microorganisms
      bacteria, 79–80
      food grade, 80–81
      fungi, 80
      microalgae, 78–79
      pharmaceutical properties, 77–78
      soil microbiota, 77
   proteins, 81–82
   research, 76
Nutrient agar, 21–23

O
Organic solvent extraction, 63, 104–107
Oscillatoria agardhii, 102

P
PAD, *see* Pulse amperometric detection
p-AMBA derivatization, 49
Paper chromatography (PC), 47, 108
Paramecium, 82
Particulate organic matter (POM), 48–49
PC, *see* Paper chromatography
PEF, *see* Pulsed electric field
Penicillium notatum, 5
Pentosans, 39
Peptides

composition determination, 91–92
extraction methods
   ammonium sulfate precipitation, 84–85
   extraction at low pH, 85
   hydrophobic resins, 86
   pH-dependent adsorption–desorption
      process, 85
   solvent extraction, 84
food proteins and, 81–82
isolation, 83
purification
   gel permeation, 88–89
   ion exchange chromatography, 86–88
   RP-HPLC, 88–90
quantification, 91–92
screening, 83
tricine-SDS-PAGE, 90–91
zymography, 91
Phaeodactylum tricornutum, 65, 112, 114
pH-dependent adsorption-desorption
   process, 85
Phenol–sulfuric acid (PSA) method, 46–47
Phormidium sp., 113, 119, 121
Phorphyridium aerugineum, 102
PHWE, *see* Pressurized hot water extraction
Phycobiliproteins, 102, 118–122
   APC (bluish-green), 102, 118–119, 121
   PC (blue), 102, 118–119, 122
   PE (purple), 102, 118–119, 121
   PEC (orange), 102, 118
Pigments, 26
   extraction and purification
      carotenoid, 108–118
      chlorophyll, 104–108
      MAE, 104
      PEF, 103–104
      phycobiliproteins, 118–122
   features
      carotenoids, 101–102
      chlorophyll, 101
      phycobiliproteins, 102
Plasma desorption, 91
PLPW, *see* Pressurized low-polarity water
   extraction
Polyunsaturated fatty acids (PUFA), 25–26
Prebiotics, 129–130, 138
Precipitation, 43, 84–86, 119
Pressurized fluid extraction, *see* Pressurized
   liquid extraction (PLE)
Pressurized fluid extraction (PFE), 45, 108
   of carotenoids, 113–115
Pressurized hot water extraction
   (PHWE), 45

Pressurized liquid extraction (PLE), 44–45, 113–115
Pressurized low-polarity water extraction (PLPW), 45
Primary carotenoids, 102
Probiotics
    applications
        CLA, 139
        clinical, 138
        food products, 138
        microbial infections, 139
    aquaculture
        crustaceans, 134–135
        disease management, 132
        fish eggs and larvae, 133–134
        fish juveniles and adults, 134
        mollusks, 135
        probiotic preparation, 132–133
        strains, 133
        use, 132
    definition, 129–130, 132
    development, 136–137
    future directions, 139
    larviculture, 135–136
    mechanism of action, 131
    nanotechnology applications, 137–138
    production, 131
    property, 130
    public health, 136
    therapeutic applications, 130
    in vitro screening, 131
    in vivo studies, 131
Probiotics, marine, 26–27
Protein hydrolysates, 81
Proteins, 7–8, 37, 41–44, 63, 81–82, 84–86, 88, 90
Prowl, 92
Psychrophiles, 21–22
Pulse amperometric detection (PAD), 48–49
Pulsed electric field (PEF), 103–104
Pyrolobus fumarii, 21

Q
Quadrupole time of flight (Q-Tof), 91–92
Quantification, carbohydrates
    chromatography
        enzymatic methods, 50
        GC-FID, 49–50
        HPAEC with PAD, 48–49
        HPLC, 48
        IEF, 51
        low-pressure liquid chromatography, 48
        mass spectrometry, 50

PC, 47
    reversed-phase high-pressure liquid, 49
    TLC, 47–48
    colorimetric/spectrophotometric method, 46–47

R
Reverse-phase high-performance liquid chromatography (RP-HPLC), 49, 88–90
Rose Bengal agar, 22
RP-HPLC, *see* Reverse-phase high-performance liquid chromatography

S
Salmonella typhimurium, 79, 138
Sample collection, 20–21
Scenedesmus sp., 63–64, 66, 68–69, 82, 106, 108, 112, 116, 118
Screening, bioactive compounds
    biomedical activities
        antibacterial activity, 27–28
        anticancer activity, 28–29
        antifungal activity, 28
        anti-HIV activity, 28
        antioxidant activity, 29
    industrial activities
        biosurfactants, 25
        enzyme inhibitors, 24
        enzymes, 23–24
        exopolysaccharides, 25
        marine probiotics, 26–27
        pigments, 26
        PUFA, 25–26
SCUBA diving, 21
Seaweeds, 7–8
Secondary carotenoids, 102
Secondary ion MS, 91
Single-dimension gel electrophoresis, 43–44
Size exclusion chromatography, 43, 88
Sodium dodecyl sulfate polyacrylamide gel electrophoresis (SDS-PAGE), 44
Soil microbiota, 77
Solvent extraction, 44–45, 63, 84, 104–108, 112–113, 118, 120–121
Spectrophotometry, 46
Spirulina sp., 82, 101, 118–120
Sponges, 4–6, 8–9, 11, 21, 29, 77, 79
Staphylococcus aureus, 9, 79, 83
Starch, 39, 41, 47; *see also* Carbohydrates
Starch agar plates, 23
Streptomyces sp., 11, 26, 79, 82
Supercritical fluid extraction (SFE), 44, 107

Supercritical fluid extraction with CO$_2$ (SCCO$_2$), 107, 115–116, 118
Superheated water extraction (SHWE), 45
Synechococcus sp., 107

T
Thermophiles, 21–22
Thin-layer chromatography (TLC), 47–48, 108
TLC, *see* Thin-layer chromatography
trans-β-carotene, 117
Transesterification, 61–62, 66
Tricine-SDS-PAGE, 90–91
Trioxacarcins, 11
Triple quadrupole ion trap, 91
2,4,6-tripyridyls-triazine (TPTZ), 40, 47

U
Ultrasound-assisted extraction (UAE), 45

V
Vacuum-microwave-assisted extraction (VMAE), 104
Verrucosispora strain, 9
Vibrio sp., 79, 133–135, 138–139
VMAE, *see* Vacuum-microwave-assisted extraction

W
Wet biomass, 62, 66

X
Xanthophyllomyces dendrorhous, 104
Xanthophylls, 102

Z
Zeaxanthin, 113–114, 116–118
Zymography, 91